1930s & 1940s

The second generation of four Gehl brothers are the sons of John W. Gehl.

Left to right: Dick Gehl, Mark Gehl, Mike Gehl (one of the first generation of Gehl Brothers), Al Gehl, and Carl Gehl.

Gehl continues to grow in the farm equipment industry, becoming the first to introduce mass-produced field forage harvesters, self-propelled forage harvesters, multiple-row crop attachments, six-foot flail choppers, silo-based re-cutters, and on-the-farm portable feedmakers.

1950s

1960s

Market Changes

1967 Gehl Brothers Manufacturing Company changes its name to Gehl Company.

1973 Gehl manufactures skid loaders and self-propelled machines. Gehl forms a marketing subsidiary in West Germany to sell the Gehl line in Europe, the Middle East, and Africa. Gehl opens a plant in Madison, South Dakota, to produce skid loaders and round balers.

1975 CB 800, the largest pull-type forage harvester, is created.

1976 Totally hydraulic grinder-mixer is produced.

1977 Gehl produces the CB 1200, the largest forage harvester in the industry. Design changes from a blower mounted behind the cylinder to a Gehl first – the offset blower.

1980 The United States enters an agricultural depression. Over the next five years, the number of farms and dealerships shrinks. Equipment makers consolidate, retreat to their core businesses, or go out of business.

GEHL

1970s & 1980s

Gehl – the first name in forage harvesting. Gehl introduced the first pull-type forage harvester more than 50 years ago.

Transition

1985 Gehl broadens its agricultural line, diversifies into the construction market, and expands its international sales. Gehl acquires a box spreader and the advanced Scavenger manure spreader.

1986 More hay tools are added to the line through acquisition.

1988 Dynalift® telescoping-boom forklifts are acquired. Two new forage harvesters and an expanded line of hay tools are offered. Acquisition of the mixer-feeder and feeder box make the Gehl line the industry's most complete.

1989 Gehl becomes publicly held. Two entities are created to focus on customers in agriculture and construction.

1991 Asphalt-paving equipment is added to the line of light construction equipment. The Scavenger sludge spreader is selected as one of 50 most innovative products of the year.

William D. Gehl
Chairman and CEO
Gehl Company

Gehl acquires **Mustang Manufacturing Company**, Inc. in late 1997, giving Gehl a stronger presence in one of the fastest-growing segments of the construction equipment industry – skid loaders.

MUSTANG

1990s & 2000s

THREE GENERATIONS OF SUCCESS

Gehl Company

1859–2009

G
E
H
L

THREE GENERATIONS OF SUCCESS

Gehl Company

1859–2009

by
Bill Beck

THE
DONNING COMPANY
PUBLISHERS

The Donning Company Publishers
184 Business Park Drive, Suite 206
Virginia Beach, VA 23462

Steve Mull, *General Manager*
Barbara Buchanan, *Office Manager*
Heather Floyd, *Editor*
Stephanie Danko, *Graphic Designer*
Derek Eley, *Imaging Artist*
Debby Dowell, *Project Research Coordinator*
Tonya Hannink, *Marketing Specialist*
Pamela Engelhard, *Marketing Advisor*

Steve Mull, *Project Director*

Library of Congress Cataloging-in-Publication Data

Beck, Bill, 1945-
Three generations of success : Gehl Company, 1859-2009 / by Bill Beck.
p. cm.
Includes bibliographical references.
ISBN 978–1–57864–524–4 (hardcover : alk. paper)
1. Gehl Company—History. 2. Agricultural machinery industry—Wisconsin—West Bend—History. 3. Construction equipment industry—Wisconsin——West Bend—History. I. Title.
HD9486.U6G443 2008
338.7'631370973—dc22
2008036270

Printed in the United States of America at Walsworth Publishing Company

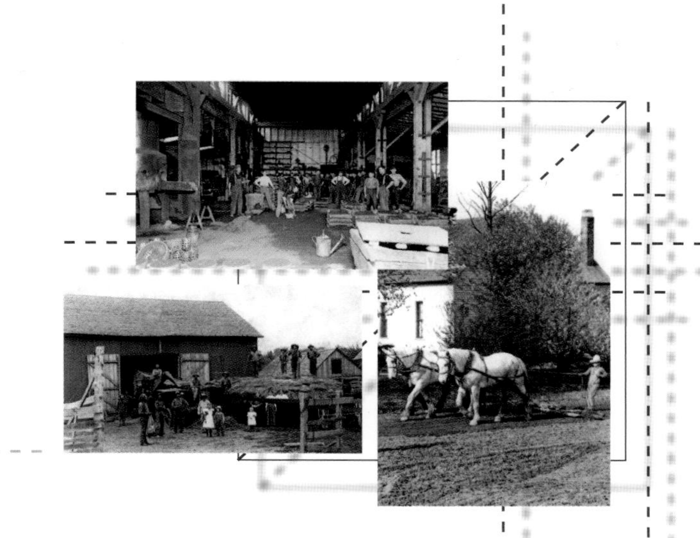

TABLE OF *Contents*

This history of Gehl Company is dedicated to the founding Gehl brothers, John, Mike, and Henry, without whose vision and determination the Company's 150 years of success would not have been possible.

Foreword

The 150th anniversary history of Gehl Company is an affirmation of the continuing strength of American manufacturing. Far from being dead, American manufacturing, as manifested by firms such as Gehl Company, is providing the world with equipment made in America. Gehl Company skid-steer loaders are as well-accepted in Europe as they are in the United States and Canada.

Gehl Company's legacy stretches back to Wisconsin in the years before the Civil War, when French immigrant Louis Lucas established a foundry in a small community at the west bend of the Milwaukee River. German immigrant Charles Silberzahn nurtured the firm's fortunes from the early 1870s to the dawn of the twentieth century. Silberzahn's invention of innovative ensilage cutters positioned his foundry to serve the growing dairy farming industry of southeast Wisconsin in the late nineteenth century.

My grandfather, Mike, and his brothers, John, Henry, and Nick Gehl, bought into the business shortly after the turn of the twentieth century and made Gehl Brothers Manufacturing Company a symbol of quality in the American agricultural implement manufacturing industry. Gehl forage harvesters, grinder mixers, and other agricultural implements helped generations of American dairy and livestock farmers efficiently manage tasks around the farm that had previously been among the most highly labor-intensive tasks in all of American agriculture.

Gehl Brothers prospered during the boom time for agriculture in the years immediately following World War II. The energy crises of the early 1970s pointed out the wisdom of diversification, and Gehl Company diversified during the decade into compact construction equipment. During the next thirty-five years, Gehl skid-steer loaders and telescopic handlers became a standard for quality for consumers in Europe and North America.

The growth of Gehl Company's construction equipment business was sadly accompanied by a corresponding decline in the Company's agricultural implement business. In 2006, Gehl Company made the decision to discontinue manufacturing agricultural implements. A weak dollar after 2005 helped Gehl Company to establish its skid-steer loaders as one of the top-selling lines in Europe.

Gehl Company's success during the past century and a half has been a direct result of the hard work and vision of the people who have made the Company their life's work. The people who have made Gehl Company what it is today reflect the work ethic of the small-town Upper Midwest. Gehl Company employees and retirees from Wisconsin and South Dakota and Gehl dealers and customers have been the rock upon which Gehl Company was built. This book is dedicated to them. It is also a tribute to our shareholders, without whose support there would not have been a 150th anniversary celebration.

While there are other manufacturing entities in Wisconsin, the Upper Midwest, and the United States that have successfully navigated the storms that have roiled the American economy for 150 years, the vision of John, Mike, Henry, and Nick Gehl and the work ethic of Gehl Company's people will ensure that the Company thrives for another 150 years.

William D. Gehl
Chairman and CEO
West Bend, Wisconsin

PLOW & CULTIVATOR
MANUFACTORY.
L. LUCAS.

Chapter One

IN THE *Beginning* 1859–1901

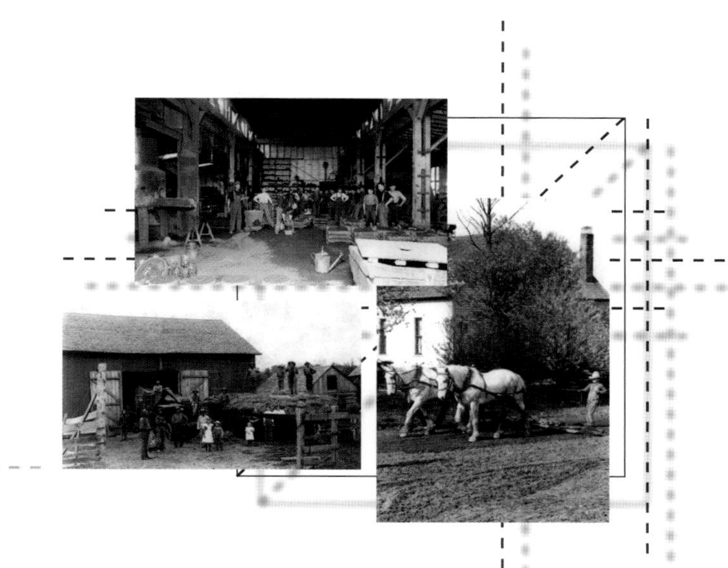

What became the Gehl Company in the twenty-first century traces its roots to a pioneer settlement at the west bend of the Milwaukee River in 1859. Wisconsin had been a state for only eleven years in 1859, and settlement was rapidly filling up the counties lying south of Green Bay on Lake Michigan, including the small farm service community of West Bend in Washington County. Population of the state more than doubled to 776,000 people between 1850 and 1860.[1] Many of the state's new residents were immigrant Germans, fleeing the political upheavals that had convulsed their native lands in 1848 and 1849. Many more were Norwegians and Swedes, fleeing the famine and landlessness in their Scandinavian homelands.

The land the new residents found was bountiful and open for settlement. The lead mining region in the southwestern part of the state ignited one of the nation's first mining booms in the 1830s, and the abundant pine forests of the northern half of the state created a wood products industry in Wisconsin that would be a mainstay of the state's economic sector more than a century and a half later. But it was agriculture that attracted the majority of the state's immigrant population to Wisconsin in the mid-nineteenth century.

King Wheat

Wisconsin in 1859 was the nation's breadbasket. Wisconsin farmers produced twenty-nine million bushels of wheat a year, more than 15 percent of the nation's total.[2] Blessed by rolling prairies and the fertile soil left behind by the retreat of the last glaciers some 10,000 years before, Wisconsin is drained by the Mississippi and Wisconsin rivers. Pioneer farmers were able to ship their wheat and corn to markets in Chicago and the East via growing ports on the Great Lakes, which border the state to the north and east.

The manufacturing sector of Wisconsin's economy, which came to the fore in the second half of the nineteenth century, was a direct result of the biggest problem that Wisconsin farmers faced in the 1840s and 1850s. As it had been for centuries, the scythe was the basic tool used in harvesting wheat. But as farmers expanded their holdings in the rich bottomlands of Wisconsin, they found that the scythe, unless wielded by an army of hired labor, wasn't up to the task of large-scale harvesting. The search for a mechanical replacement for the scythe and other agricultural implements was the incentive for the state's first industrial boom. So, too, was the need for modern weapons in the Great Civil War that was about to descend on the American Republic.

Before the mechanization of farm equipment, the scythe was the most common tool used for reaping grain.

The Lucas Foundry

Wisconsin was the cradle of the nation's agricultural implement manufacturing industry. Wisconsin inventors began patenting wheat-harvesting equipment as early as the 1840s. Jerome I. Case built his first factory in the Lake Michigan port city of Racine in 1847 to produce a mechanical thresher powered by horses.[3] In the years to come, Case Threshers, with as many as ten horses providing the motive power, became a ubiquitous part of the landscape in Wisconsin wheat fields in October. The Case Thresher allowed the farmer to quickly harvest large fields of wheat, thus avoiding the spoilage that was so prevalent with the scythe.

In 1859, Cyrus McCormick moved his reaper manufacturing company to the booming city of Chicago, just fifty miles south of the Wisconsin state line. McCormick had introduced the first modern mechanical reaper in 1831 in Virginia, revolutionizing farming in the United States. By 1860, his Chicago firm was selling 20,000 reapers a year.[4] In 1857, George Esterly built a manufacturing plant for a wheat-harvesting machine in the Village of Whitewater.[5] The agricultural implement manufacturing business flourished in a corridor between Chicago and Milwaukee in the years following 1860.

There were numerous reasons for the industrial boom in southeastern Wisconsin. The area enjoyed a ready marketplace, with thousands of wheat farmers living in the state and adjacent Illinois, Iowa, and Minnesota, all wheat-producing states themselves. Milwaukee's German immigrants were skilled mechanics who enjoyed access to capital amassed by the thrifty German immigrant community. And Milwaukee's port was able to inexpensively handle iron ore shipped down the Great Lakes from the iron ranges of nearby Michigan and northern Wisconsin.

The availability of iron ore and the growing success of the agricultural implement manufacturing industry spawned other industries in the metropolitan Milwaukee area. The need for iron castings created a thriving foundry industry in the 1850s and 1860s. American foundries had made wrought iron since before the American Revolution, mostly by heating iron ore and then removing oxygen. Beehive iron furnaces sprouted up in Pennsylvania's Lehigh Valley, upstate New York, and parts of New England during the late 1700s and early 1800s. By the 1850s, those foundries had moved west to the Great Lakes states.

For much of its first century in existence, the United States was a steel-importing nation. As late as the Civil War, U.S. steel imports, mostly from Great Britain, were greater than domestic

Hand tools were prevalent on the typical nineteenth-century farm. Often the entire family would lend a hand on the farm and in the fields. *(Reproduction from the collection of the Washington County Historical Society, #080,441)*

This photo, taken around 1890, shows Silberzahn (far left) supervising the iron workers in the foundry. On the right is the furnace used for melting iron and on the floor are the molds into which the melted iron would be poured. *(Reproduction from the collection of the Washington County Historical Society, #016,302)*

The Company's foundry, after being rebuilt following the 1906 fire. Iron-making was resurrected and included all departments of iron, lathe, and machine work needed to repair agricultural machinery. *(Reproduction from the collection of the Washington County Historical Society, #019,067)*

production. The postwar demand for iron ore in the 1870s and 1880s was fueled by the increasing demand for steel.

Henry Bessemer of Great Britain solved one of the nineteenth century's most difficult metallurgy problems in 1856 when he patented a steelmaking process named after himself. Those using the Bessemer process made steel by blowing air into molten iron ore. The resulting steel had high tensile strength, was ductile, and was suitable for shaping into many forms. By the mid-1860s, Bessemer's pear-shaped converters were quickly replacing the iron industry's beehive furnaces.[6]

One of those foundries was started in 1859 on what was then known as River Street in West Bend. Louis Lucas, who had previously worked in West Bend as a tin-plate worker and coppersmith between 1852 and 1859, built the foundry near the lower bridge on River Street, where he carried on the business until 1873.[7] Then thirty-nine years old, Lucas had been born in France of Huguenot Protestant stock. At the age of fourteen, he was apprenticed to a coppersmith in France, and later served a three-year hitch in the French Navy during the early 1840s. Lucas came to the United States in the 1850s and arrived in West Bend late in the decade.[8] Like most small-town foundries of the day, the Lucas Foundry poured castings for making and repairing agricultural implements and other hardware in general use around a farm.

At the time Lucas was beginning his business, West Bend was the thriving Washington County Seat. More than 800 people lived in the community when the federal census-takers

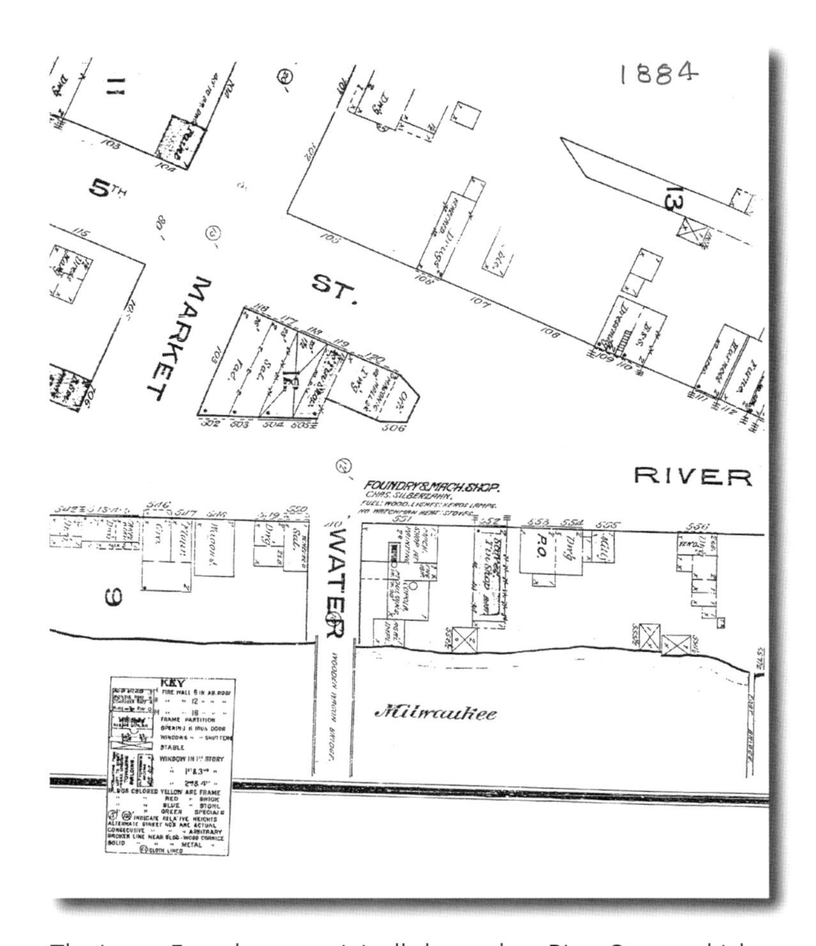

The Lucas Foundry was originally located on River Street, which is now known as Main Street. Local farmers relied on foundries such as Louis Lucas' for the repair of farm implements.

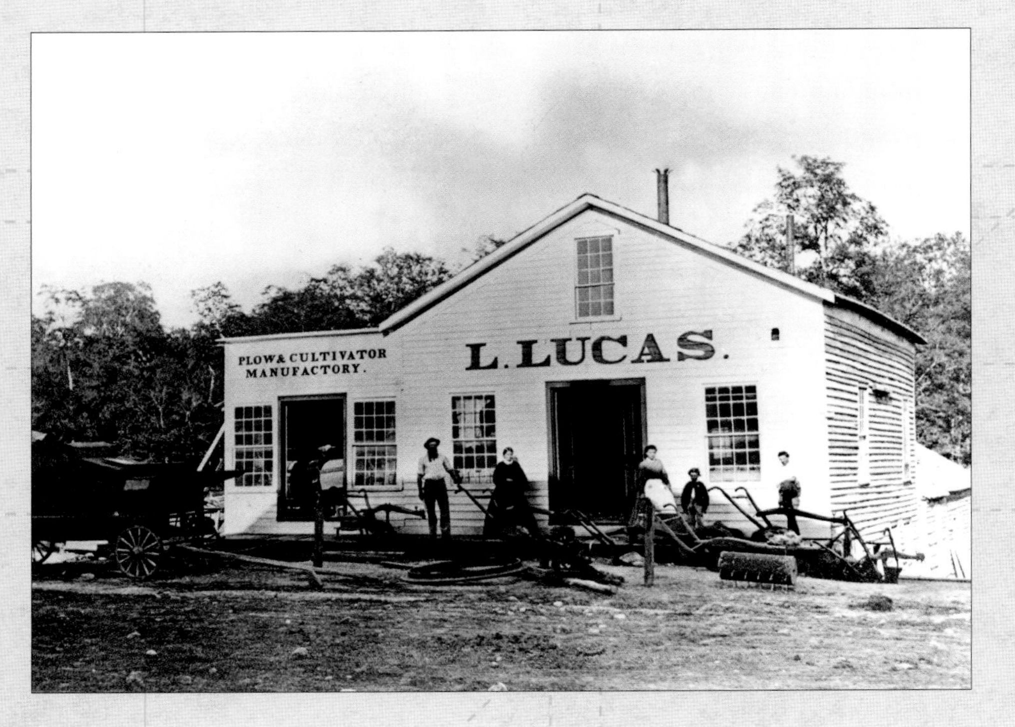

Left: The Louis Lucas Foundry on River Street. The reason Louis Lucas sold the business to Jacob Young in 1873 is unknown, but it could have been in part due to the 1873 death of Lucas' wife, Eulalie. Together they had three children: Edward, Mary, and Henry. Two other sons died at early ages. After selling the foundry, Lucas went on to operate a very successful cranberry farm in Washington County.

Middle: A late-nineteenth-century view of Water Street, looking toward the West. In the distance, the railroad, which arrived in West Bend in 1873, can be seen.

Bottom: The Milwaukee River in the nineteenth century.

The threshing machine was a staple in the farming community throughout the nineteenth century. It mechanized the process of separating grain from stalks and husks. *(Reproduction from the collection of the Washington County Historical Society, #080,476)*

The office of the *West Bend Beobachter* (German for "observers") was shared with another local newspaper, the *West Bend Democrat*. The longtime editor of the *Beobachter* was Carl Quickert. *(Photo courtesy of the West Bend Chamber of Commerce)*

made their rounds in 1860, and the community boasted several milling and lumber manufactories, a brewery, a woolen factory, three churches, a newspaper, and a school. Pioneers remembered a half-century later that "at the beginning of the Civil War, it had grown to be one of the most promising villages of the state."[9]

Founded in November 1845, West Bend's population included a large number of German immigrants. Accordingly, one of the first businesses established in the community was a brewery, the predecessor of the West Bend Brewing Company in the early twentieth century, which furnished residents with its popular "Lithia" brand beer.[10] Another longtime West Bend business was the *West Bend Beobachter*, the local German-language newspaper, which survived well into the twentieth century.[11]

Much of the land where Lucas located his foundry had been inundated by the flooding Milwaukee River in the spring of 1858. The *West Bend Democrat* noted in early June 1858 that "the oldest inhabitant does not recollect the time when the Milwaukee River was higher than it was last Saturday."[12] Flour and lumber mill dams along the course of the river were swept away by the spring flood, which no doubt created replacement castings business for Louis Lucas' new foundry the next year.

Originally, the foundry business was established to provide only for local needs; mostly implement repair. An early tintype photo of the Lucas Foundry shows one of the farm machines of the time, a threshing machine, the threshing cylinder of which is a large wooden cylinder bristling with spikes apparently driven into the wood. Other products of that time were plows, plow points, tread powers, cow stanchions, farm feed cookers and foundry products such as sleigh hardware, clothes reels, hay tedders, pump jacks, lifting jacks, manure loaders, and saw frames.[13]

The Civil War and After

In the four-year-long Civil War between 1861 and 1865, Wisconsin sent nearly 100,000 of her sons to fight to preserve the Union. Wisconsin regiments were with General Grant at Vicksburg, General George Meade at Gettysburg, and General Tecumseh Sherman on his 1864 March to the Sea. More than 12,000 Wisconsin soldiers failed to return to the Badger State after the war. The Lucas Foundry struggled to find parts and castings for its furnaces because much of the nation's iron supply was immediately consumed by the military's voracious appetite for ferrous metal. For part of the war, Lucas himself donned Union blue. Although then forty-one years old, Lucas' experience commanding a gun battery aboard a French naval ship nearly twenty years earlier had caused him to be recommended for appointment as captain of a Wisconsin militia unit.

Charles Silberzahn, shown here in uniform, moved to Wisconsin shortly after serving in the Civil War. Louis Lucas also spent time serving his country during the Civil War. *(Reproduction from the collection of the Washington County Historical Society, #016,020)*

The manpower shortage in the Wisconsin wheat fields created an ever-growing demand for agricultural implements to replace the hands of those who were away on the far-flung war fronts. Production of McCormick Reapers doubled in 1863 to more than 40,000 units as farmers in the Midwest Wheat Belt worked valiantly to feed the troops on the battlefield and civilians on the home front.[14]

In the years after the war, Wisconsin's farmers struggled as the nation's Wheat Belt moved west to the seemingly endless prairies of Minnesota, the Dakotas, and Montana Territory. Superior, at the state's far northwestern edge, and its neighbor across the bay, Duluth, Minnesota, became the Great Lakes terminus for the Great Northern and Northern Pacific railways. In the 1870s and 1880s, the two transcontinental rail lines opened up the Northern Great Plains to settlement and small grains agriculture.

The hard red spring wheat of Minnesota and the Dakotas displaced much of the winter wheat that had been grown in Wisconsin and the Midwest. Railroads hauled the flour to the mills in Minneapolis, or to the Head of the Lakes, where the spring wheat was shipped down the Lakes to the huge flour mills that dotted the Buffalo, New York waterfront. Hard red spring wheat was milled into a pure white flour that became the standard for baked products for the next 130 years. It also had another quality that recommended it to bakers. Because the spring wheat of Minnesota and the Dakotas had a higher gluten, or protein, content than winter wheat, it produced more loaves of bread per pound of flour.[15]

Demand for winter wheat grew at a breakneck pace in the late 1870s and early 1880s. By 1881, the Northern Pacific was shipping nearly ten million bushels of wheat from Superior to Buffalo each year.[16] Depletion of the soil in southern Wisconsin only aggravated the competitive problem for the state's farmers. By the mid-1870s, it was becoming more and more apparent that Wisconsin's farmers would have to diversify in order to survive.

New Owners

In the years after the Civil War, Louis Lucas employed as many as five workers in the foundry on River Street, making and repairing agricultural implements, including fanning mills, plows, and rotating churns. He was involved in the life of the community, serving two years as justice of the peace and a year as a West Bend village clerk. Lucas sold the business in 1873 to Jacob Young, who conducted the business in partnership with John Kunz and others. At that time, the Lucas Foundry advertised that "the scope of the business has been enlarged and embraces all departments of iron, lathe, and machine work required to repair agricultural machinery."[17]

It is not known why Lucas sold his foundry to Young and Kunz, but 1873 was a panic year, and businesses across the state of Wisconsin and much of the nation failed as credit dried up. The failure of Jay Cooke & Company, the principal backer of the Northern Pacific Railway, was felt particularly keenly in Wisconsin. In September 1873, Cooke's Philadelphia investment banking firm defaulted on several millions of dollars in

The inside of the Silberzahn Manufacturing Company, taken around 1890, after the new plant was constructed on Water Street. *(Reproduction from the collection of the Washington County Historical Society, #019,065)*

Northern Pacific bonds, and Wall Street and the nation's banking system all but collapsed.[18] The "Long Depression" that followed affected U.S. farmers until nearly the end of the 1870s.

For Washington County and Wisconsin, the 1870s were difficult years, as farmers diversified into a new model of agricultural productivity. The stagnating local agricultural economy, crippled by the transition from a wheat-based commodity structure, searched for a profitable replacement. In the 1870s, Yankee economists and German immigrants began preaching the benefits of a dairy-based economy for Wisconsin agriculture.

By the early 1870s, nearly two-thirds of the residents of the Upper Midwest states of Wisconsin, Iowa, and Minnesota lived on farms in the Upper Mississippi Valley. Many were turning to dairy cattle as their livelihood. They raised European brands of cattle and they sold what they could ship to consumers in Milwaukee, Chicago, Minneapolis-St. Paul, and other urban areas in the region. Dairy products at the time typically consisted of cheese and sour cream butter.[19]

Cheese-making in Wisconsin dates to the late 1830s, and German farmers were adept at making the European cheeses that a nation of immigrants in America's cities demanded for their dinner tables. At a time when refrigeration was all but nonexistent, butter was easier to handle than liquid milk, and the number of creameries in Wisconsin increased dramatically between 1875 and 1900. The University of Wisconsin and its Extension program actively promoted the industry during the last years of the nineteenth century, and the College of Agriculture worked with dairy farmers statewide on such techniques as increasing butterfat content of milk and testing for bacteria in milk that led to breakthroughs in pasteurization.[20]

William Dempster Hoard tirelessly promoted the industry in the pages of his weekly magazine, *Hoard's Dairyman*. In 1872, Hoard helped build the first cheese factory in Jefferson County and helped found the Wisconsin Dairymen's Association.[21] Hoard was so well known for his championing of the state's dairy industry that he rode that popularity to election as Wisconsin's sixteenth governor, serving from 1889 to 1891.[22] By the time Hoard left the governor's mansion in Madison, nearly 90 percent of the state's farms raised dairy cows.

{ DAWN TO DUSK DRUDGERY }

Dairy farmers in Wisconsin in the late nineteenth century dealt with drudgery from dawn to dusk every day of their lives. Tending dairy cattle was very labor-intensive. Farmers and their families were up with the sun. They had to water and feed the cows every morning.

Farms were small. The typical dairy farm in the state in the late nineteenth century consisted of one to two dozen milkers. The cows had to be milked by hand twice a day. The milk and cream had to be hauled to the nearest town or creamery on a regular basis, or it would spoil. If the dairy farmer was fortunate enough to be within a day's distance of a local cheese factory, he had a ready-made market for his milk.

Cows were milked from spring to late fall. Milk was poured into earthen jars and stored in a cellar below ground. Its storage in a cool place allowed the cream to rise to the top, where it was then skimmed off with a wooden spoon. The cream was churned into butter and sold at a local store.[28]

Keeping animals fed was a major part of farmers' chores. Land had to be prepared and planted, often in corn. Preparing the land meant ditching it so it would drain. By the late nineteenth century, most of the southern half of the state had been cleared, so the arduous task of burning and removing logs and brush had been performed a generation or more before. But land that had not been tilled in some years had to be torn up with a horse-drawn drag, often homemade and owned in common with several other local farmers.

Implements for dairy farms were still fairly primitive in the 1880s and 1890s. One dairy farmer from nearby Two Creeks noted that dairy farmers at the time relied on "the axe, crosscut saw, hoe, shovel, plow, and drag" to prepare land for tilling and planting.[29] The introduction of the Hexelbank Cutter in 1889 promised to make at least the preparation of feed for the cows less of a labor-intensive task.

For the dairy farmer's wife, there were relatively few labor-saving devices in the nineteenth century. Water for bathing and drinking—for both humans and cows—was pumped by hand. The wife of a Wisconsin dairy farmer cooked meals on a wood-fired stove.[30] Kerosene lamps lit farmhouses and barns; the danger of fire was uppermost in every farmer's mind.

Nineteenth-century farm implements were pulled by teams of horses. Here, a farmer prepares the seed bed in his corn field.

Washing and ironing clothes for the family was perhaps the worst job the wife of a dairy farmer faced each week. Two or three tubs of soapy and rinse water were filled and heated on the stove in the kitchen. A hand-cranked wringer attachment on one of the tubs allowed the farmwife to remove excess water. Ten-pound flat irons were heated on the stove and used to iron the family's shirts, pants, and dresses. Farm wives called them "sad irons," because they often sustained burns from the red-hot metal.[31]

It wouldn't be until the introduction of farm mechanization in the 1910s and 1920s and electric power in the 1930s that the workload of the Wisconsin dairy farmer and his family finally lessened. By that time, corn had been supplanted as silage by clover and alfalfa, and the need for a Hexelbank Ensilage Cutter had passed. But Charles Silberzahn's 1889 invention could well be considered one of the first steps in the mechanization of America's dairy farms.

The Hexelbank Cutter

The post-Civil War transition from wheat to corn and dairy farming in Wisconsin and the Upper Midwest was perhaps the state's most important agricultural development in the nineteenth century. Dairy farming, however, was hard work. Cows had to be fed, bedded in clean straw, milked, and herded. There was no free time for the dairy farmer, as there was between planting and harvesting for the wheat farmer. And dairy farmers had to tend to a crop of their own; planting corn to use as feed for the stock.

Cutting the corn stalks was a labor-intensive, grueling task that involved the entire farm family. The stalks had to be gathered, bundled in twine, and loaded into wagons for the trip to the barn. The farmer and his older children used sharp knives to chop the stalks into silage for the cattle; the younger children turned grinding wheels so the farmer and his older children and hired hands could keep their knives sharp for cutting the silage.

In 1878, Jacob Young and John Kunz brought Charles Silberzahn into the Lucas Foundry as a partner. Silberzahn had come to West Bend from nearby Sheboygan, where he had partnered with John Michael Kohler in founding the Kohler and Silberzahn Foundry, an iron foundry on the shores of Lake Michigan. Kohler and Silberzahn pooled the then-immense sum of $5,000 and bought the business from Kohler's father-in-law, Jacob Vollrath.[23]

Silberzahn was a native of the German Grand Duchy of Baden. Born in 1828, he was twenty years old when a revolution swept the German principalities in 1848. Drafted into the artillery and ordered to fire on fellow countrymen, he fled to Switzerland and eventually to the United States. He came to St. Louis in the 1850s, and worked on riverboats on the Mississippi and Illinois rivers. Silberzahn found himself in Memphis when the Civil War broke out. He did repair work for the Union riverboat fleet in the wake of the Battle of Vicksburg in 1863, and helped escort prisoners and wounded north following the battle.[24] Following the war, Silberzahn moved first to Dubuque, Iowa, and then to LaCrosse. In the early 1870s, he lived in Milwaukee, where he spent two years as foreman in the machine shop of E. P. Allis.[25]

Mr. Charles Silberzahn, inventor of "The Light Running Silberzahn" Ensilage Cutter.

The Kohler and Silberzahn Foundry in Sheboygan after the Civil War specialized in the manufacture of agricultural

Charles Silberzahn, pictured here at the age of ninety, was born in 1828 and died in 1921. Before joining the Lucas Foundry, Silberzahn was co-owner of a Sheboygan iron foundry that manufactured and repaired products similar to Lucas'.

West Bend
FOUNDRY
and
MACHINE SHOP
Silberzahn & Jung
Sole Manufacturer of celebrated
CHAMPION FEED CUTTER
The best in the market. Two sizes manufactured.
Also a large supply of Plows, Iron Kettles
Small castings of all kinds kept on hand and made to order on short notice
Highest prices paid for old Iron

The Silberzahn Manufacturing Company produced a variety of goods besides farm implements. Pictured here are cast-iron hitching posts manufactured by the company.

In this photo taken in approximately 1882, Charles Silberzahn (right) is pictured here with August Bernhagen, the Silberzahn plant engineer at the time, standing next to the steam engine that powered the plant. *(Photo courtesy of* The West Bend Daily News*)*

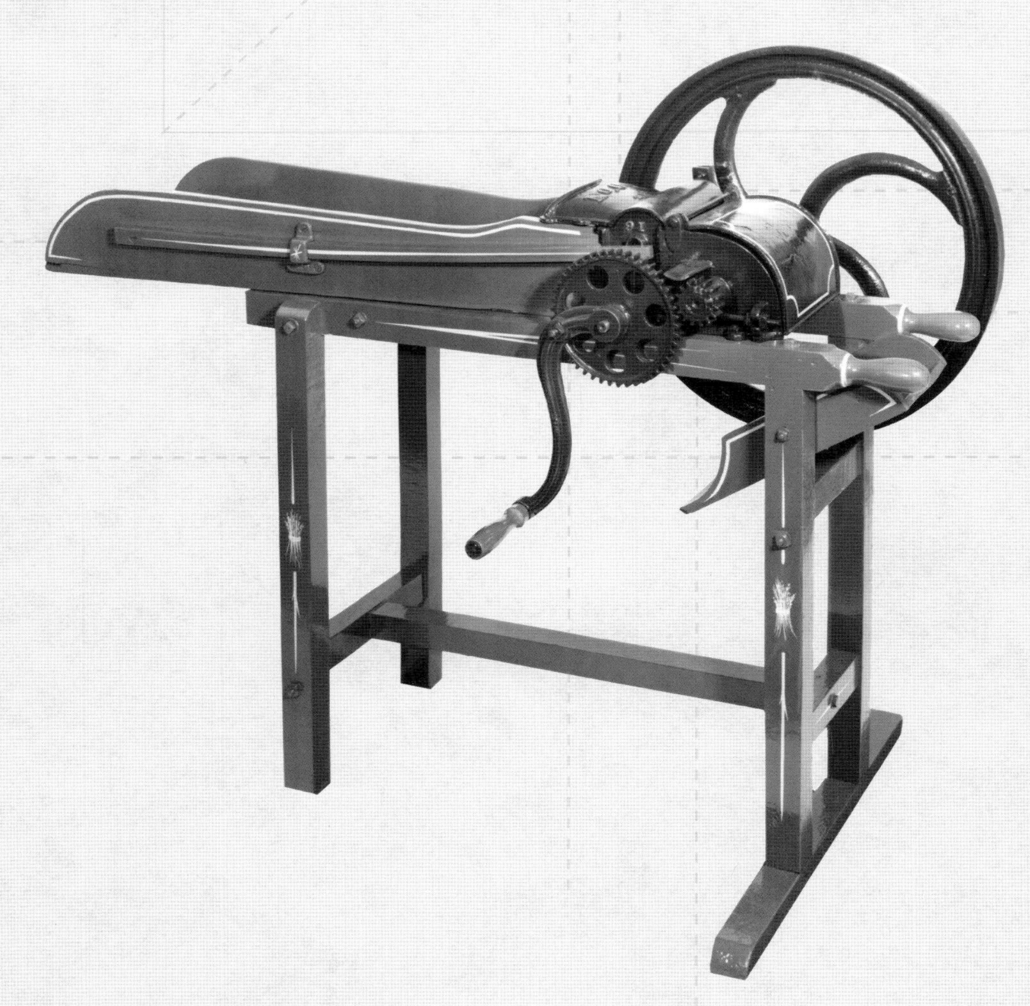

The Hexelbank served the needs of livestock farmers from 1880 to 1906.

The Silberzahn Manufacturing Company plant, as it appeared after being built on its new Water Street site. This building would later succumb to a fire in 1906. Charles Silberzahn is easily identified by his long, white beard.

implements, including water troughs and a scalding vat for removing the hide from livestock carcasses. In 1878, Silberzahn sold his interest in the foundry to Herman Hayssen and John H. Steyn, took his proceeds, and purchased a partnership in the Lucas Foundry. Kohler and his new partners expanded the Sheboygan foundry in 1880, but lost their entire investment just months later when the business burned down. Kohler and his partners rebuilt on the site and in 1883 introduced a line of enameled plumbing fixtures.[26] Kohler Company went on to become one of Wisconsin's most successful manufacturers.

Although the roots of the Company's farm implement business lie in the hand-cranked Hexelbank, it often goes unnoticed as compared to its successor, the Silberzahn. Literally translated, the German word "hexelbank" means a cutting or slicing bench.

Silberzahn, meanwhile, took his mechanical expertise to West Bend and began tinkering with a machine he hoped would revolutionize feed cutting for dairy cattle. Silberzahn had produced a working model of what was to become known as the Hexelbank Cutter during his years in Sheboygan. In 1889, Silberzahn and the Lucas Foundry introduced the first commercially produced Hexelbank Cutter. It was a cylinder-type, hand-cranked feed cutter used for cutting corn for livestock. The unique machine was initially sold mostly in Wisconsin, but quickly found favor with dairy farmers in the rest of the United States. The Hexelbank Cutter sold at that time for about $11.50.[27]

The introduction of silos into Wisconsin's agricultural economy allowed dairy farmers to store livestock feed, or silage, safely and economically. Cutting feed mechanically for the silos led to the introduction of the Silberzahn Ensilage Cutter with its patented reversible gear. Silberzahn also began production of feed elevators for filling silos and upper stories of barns.

In 1890, Silberzahn and his sons bought out some of the original investors in the Lucas Foundry, incorporated it as the Silberzahn Manufacturing Company, and concentrated on the manufacture of ensilage feed cutters. The antecedent of the modern Gehl Company was now in place.

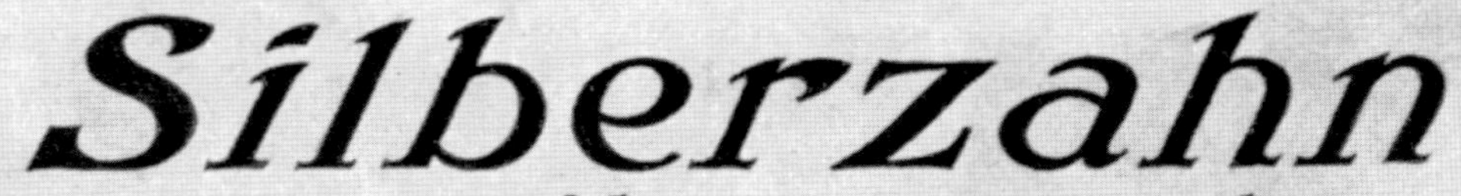

Ensilage and
Feed Cutters

Manufactured By

Gehl Bros. Mfg. Co.

West Bend, Wis.

Established 1879 Incorporated 1890

"The
King
Of
Ensilage
Cutters"

Chapter Two

GEHL BROTHERS *Manufacturing* COMPANY 1902–1919

In the 1890s and early 1900s, Silberzahn's ensilage feed cutters had legions of users on dairy farms in Wisconsin and surrounding states. By the 1910s, Gehl Brothers Manufacturing Company, as the Company was later renamed, was marketing the Silberzahn Ensilage Cutter across a broad swath of the American Farm Belt. Sales after 1910 were especially strong in the prairie states of North and South Dakota, Nebraska, and Kansas.

One Kansas farmer wrote the Company in 1913 that he and his sons had been chopping silage with their Silberzahn since 1911. "We have used it in the past two seasons," he said, "the first season using it to fill three silos, and last season six silos, and it has given perfect satisfaction. Have needed no repairs on it in the two years."[32]

A. M. Johnson, who milked a herd of Holstein cattle near Doniphan, Nebraska, noted that he had sold a competing ensilage cutter and "bought a No. 18 'Silberzahn.' After cutting six hundred ton, I can say it is the best cutter I have ever seen."[33] Professor C. F. Curtis of Iowa State College at Ames enclosed a photograph of the Department of Agriculture's Silberzahn Cutter in action on the college's farm. "The machine gave good satisfaction," Curtis said. "I have also used one on my own farm with good results."[34]

Dairy farmers across the Midwest knew and trusted the name Silberzahn. They didn't yet know much about the Gehl brothers.

The Gehl Family Arrives in Wisconsin

The Gehl family first came to Wisconsin in 1845. They were among the millions of new Americans who came to the United States between 1840 and the end of the nineteenth century, attracted to their new homeland by the prospect of freedom.

Sometime in the spring of 1845, Mathias Gehl and his wife, Margaret Berscheid, left their home village of Keispel, Luxembourg, for America. Gehl, thirty-five, had apprenticed as an iron forger in nearby France. When he arrived in New York, he heard good things from German-speaking compatriots about the opportunities that awaited them in the small community of Milwaukee. Mathias Gehl and Margaret packed up their children and booked passage for Wisconsin.[35] The family name then was spelled "Geehl," but it was pronounced the same as it is today. Mathias' son Michael dropped the second *e* in the early 1890s.

The Luxembourg they left in 1845 was still recovering from the Napoleonic Wars, which had ended thirty years before. For much of the 1830s, the Duchy of Luxembourg had been occupied by Prussian troops, and in 1839, the First Treaty of London had partitioned the small country between French-speaking Belgium and the German Confederation to the East.

In 1842, the Grand Duchy of Luxembourg furthered its ties to Germany when it became part of the German Customs Union.[36]

The political instability of the region and overcrowding in the tiny country had convinced Mathias and Margaret that their destinies laid elsewhere. The young couple was accompanied to the United States by several adult siblings. Other Gehl family members in Luxembourg emigrated to the Austro-Hungarian Empire and to neighboring Belgium.[37]

When Mathias arrived in Milwaukee in 1845, he was appalled at what he found. What had been Solomon Juneau's fur post not very many years earlier was now a growing city of 20,000 people. More than a dozen flour mills dotted the city's waterfront, and a cacophony of ship's whistles and horns split the air from morning until night. Milwaukee's port was the terminus of the Wheat Belt of southern Wisconsin. The city's meatpacking industry was growing quickly in 1845, and the procession of cattle and hogs being herded through the streets to slaughter left Milwaukee's avenues ankle-deep in offal.[38]

Mathias Gehl had immediately managed to secure work upon his arrival in Milwaukee. The City of Milwaukee offered to build a shop for the newly arrived immigrant if he would agree to stay and do the iron forging for Milwaukee and the surrounding area. There was a great demand for draw shaves, kitchen knives and utensils, blacksmith tools, and farm implements.

Gehl, however, was less than impressed with the living conditions. "In a mudhole such as this, I won't stay for a living," Mathias Gehl proclaimed to Margaret and their children.[39] Mathias and Margaret packed up their children and a few belongings and set out on foot along the Milwaukee-Green Bay Trail. They got about thirty-five miles north of downtown Milwaukee when they arrived at the small community of St. Lawrence, just inside the Washington County line.

Like many immigrants, Gehl had purchased land at the Government Land Office in Milwaukee. The Gehl homestead was located two miles west of St. Lawrence on forestland in Section Four of the Town of Hartford. Gehl paid $1.25 an acre for the land.

The Silberzahn at the Iowa State College in Ames, Iowa.

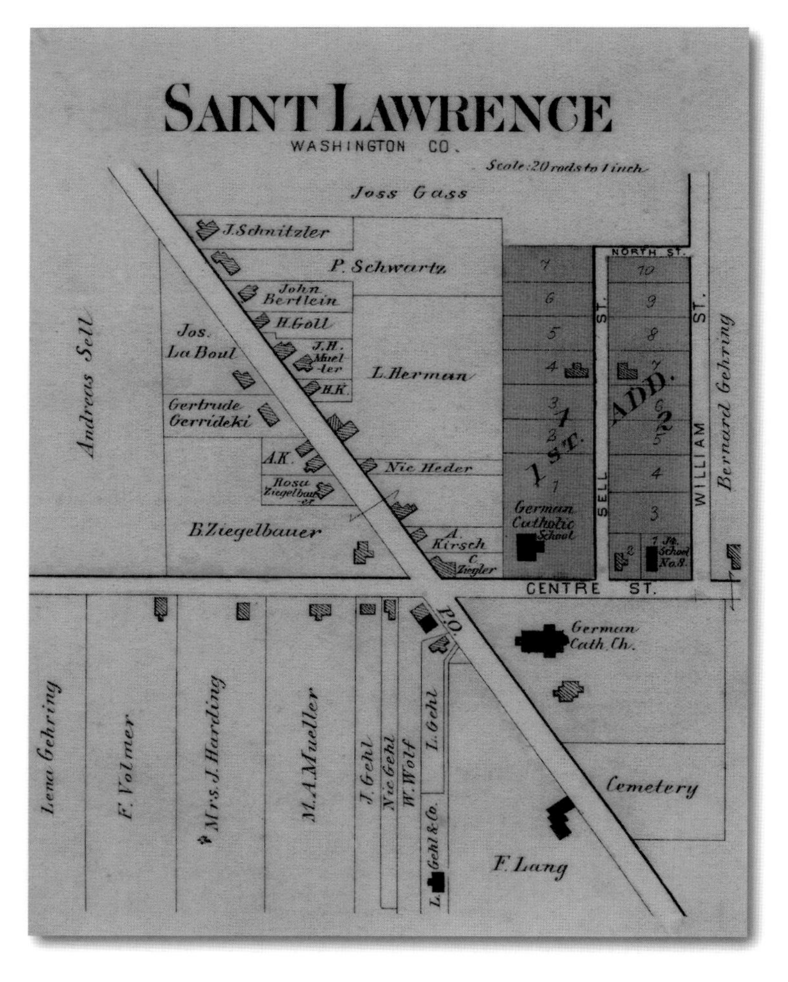

An 1892 map of the St. Lawrence region, which the Mathias and Margaret Gehl family inhabited upon their arrival in Wisconsin. The owners of certain properties in the community are shown on the map, including land owned by Gehl family members. *(Courtesy of the Washington County Historical Society)*

Immediately after arriving on the new land, Mathias, Margaret, and some of their neighbors cleared trees from the site and built a small cabin. During the ensuing winter, Gehl and his neighbors continued to clear the land, and the next building erected was a small forge shop. As one of the first iron-forgers and wagon-makers in Washington County, Mathias Gehl was an honored resident. In later years, he often talked about walking the thirty-five miles to Milwaukee in those early years to get flour and other supplies for his growing family.

The Gehl Family in Washington County

Mathias and Margaret Gehl had nine children. Their oldest son, Michel ("Michael"), was born in 1839 and was about six years old when the family moved to Wisconsin. Michael grew to manhood on the family farm near St. Lawrence and married Theresa Netzinger in November 1861 at St. Lawrence Church in St. Lawrence. Theresa Netzinger, three years younger than her husband, was also born in Luxembourg, in 1842, and accompanied her parents to America in 1852.[40] The newlyweds moved to a farm in the Town of Addison, southwest of Allenton, where they resided until 1871. Margaret Gehl died that year, and Michael and Theresa moved to the family farm near St. Lawrence to help Mathias Gehl manage the property.[41]

Michael and Theresa had twelve children, two of whom died in infancy. Four of Michael and Theresa Gehl's sons became involved in the firm that would bear the family name in 1904. Nicolaus, one of the co-founders of the Company, was born in June 1868 and died in Denver, Colorado, in February 1928. Johan ("John"), born in December 1872, was the longtime secretary of Gehl Brothers and then served as chairman of the Board until his death in November 1951. Michael, who was born in April 1883, had served as the longtime treasurer of Gehl Brothers and then succeeded John as chairman of the Board until his own death in 1969. Henrich ("Henry"), born in February 1885, was the last of Michael and Theresa Gehl's children to die. He served as president of Gehl Brothers Manufacturing Company for more than a half-century and became chairman of the Board upon Michael's death.[42]

Mathias Gehl died in 1896, twenty-five years after the death of his beloved Margaret. He was representative of his generation

An early photograph of the family of Michael and Theresa (nee Netzinger) Gehl. Left to right, bottom row: Jacob, Anna, Theresa, Mary, Henry, and Michael, Jr. Top row: John, Peter, Michael, Sr., Theresa (nee Netzinger), Mathias, and Nicolaus.

of immigrants: hardworking, proud of his adopted country, and centered on his family and his church. Had he lived another five years, he would have taken great pride in his grandsons' purchase of the Silberzahn Manufacturing Company in nearby West Bend.

Changing of the Guard

The company that Charles Silberzahn and his sons had purchased in 1878 and incorporated in 1890 had made the transition from a small-town Wisconsin foundry to a regional agricultural implement manufacturer. Sometime prior to 1892, Silberzahn moved from the River Street location to Water Street, which at the time dead-ended where Silberzahn Manufacturing Company relocated. The Company would ultimately remain on Water Street for more than one hundred years.

By 1900, the West Bend manufacturer was known to dairy farmers across the region for the Silberzahn Feed Cutter, one of the best machines on the market. "We have made the manufacture of feed cutters a specialty for the past years and sold thousands of them," the Company said in its 1900 catalog. "As the demand for cutters increased and farmers required larger and heavier cutters, we again found it necessary to improve upon the old-style cutters and bring them ahead of the inventions and improvements of the present day. This we have very successfully done."[43]

Silberzahn Manufacturing Company advertised that "we use nothing but the very best steel in our knives; they hold an excellent cutting edge and cannot be beaten."[44] Silberzahn stood behind its cutters, noting that "they are warranted first class in every respect, satisfaction guaranteed, and we do not hesitate in recommending them to the public as the best cutter ever offered for sale."[45]

The Company offered a half-dozen feed cutter models in 1900, ranging in price from $32 to $125. Accessories included a straight and angled delivery carrier for lifting silage into barn lofts. By 1900, the Company had diversified into producing other implements in addition to the feed cutter models. Silberzahn produced a six-foot hay tedder, a circular wood saw, a double-pinion six-horse power, a two-horse tread-power machine, pulley jacks, emery wheel grinders, corn shellers, tank pumps, anti-friction clothes reels, galvanized steel tanks, and a full line of spare parts for agricultural equipment.[46]

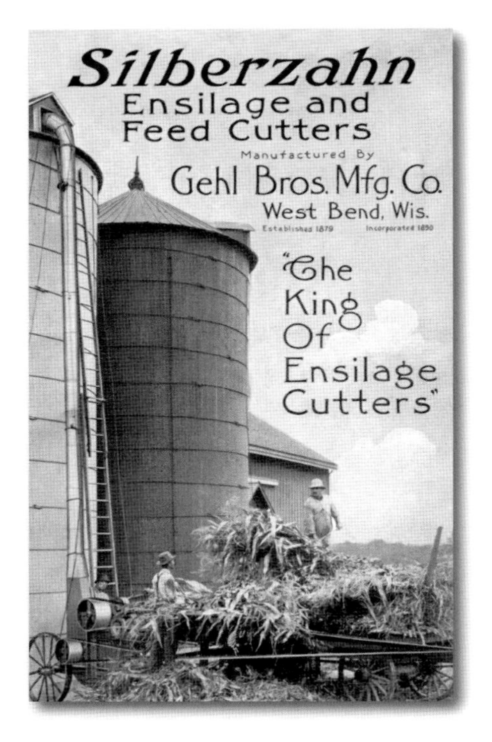

Through the collection and publication of farmers' testimonials boasting of the quality and efficiency of the Silberzahn, in 1916 the Company did not hesitate to proclaim the Siberzahn "The King of Ensilage Cutters."

The buildings in the center right are the Silberzahn plant as it appeared in 1905. In the center is the West Bend Lumber Company. *(Reproduction from the collection of the Washington County Historical Society, #016,073)*

The original line of Silberzahn cutters manufactured by Gehl Brothers Manufacturing Company.

By the turn of the twentieth century, Charles Silberzahn had been in the agricultural implement manufacturing business for some thirty years. He had also called upon his sons to be a part of the Company. Although one of his sons, Charles, Jr., died in 1901 at the young age of forty-seven, Silberzahn's other two sons, Homer and Louis, served as officers of the corporation until early 1902. Then seventy-three, Charles Silberzahn decided in 1902 that it was time to sell the business. A photograph from the 1900 Silberzahn catalog reveals a man with a long white beard wearing a campaign hat from the Civil War, in which he had served as a naval engineer.

It is said that age prompted Charles Silberzahn to sell his interests in the Company; however, even after the Gehl brothers took over, Silberzahn remained a part of the management team. In this 1912 group photo, Silberzahn is shown seated in the center. Seated around Silberzahn, from left to right, are John W. Gehl, Henry Gehl, Mike Gehl, and Nick Gehl. The gentlemen standing were the sales force at the time.

John W. Gehl had Americanized his birth name of Johan. He had grown up on the family farm near St. Lawrence and was just shy of thirty years old when he bought into the Silberzahn Manufacturing Company. John Gehl was already well known in Washington County. He married Mary Thoma of an old West Bend German family in 1900 and worked as an insurance agent and stave-maker in the Washington County Seat. John W. Gehl was active in Democratic Party politics in the county, and in 1902, he was elected to his first of four terms as Washington County register of deeds.[47]

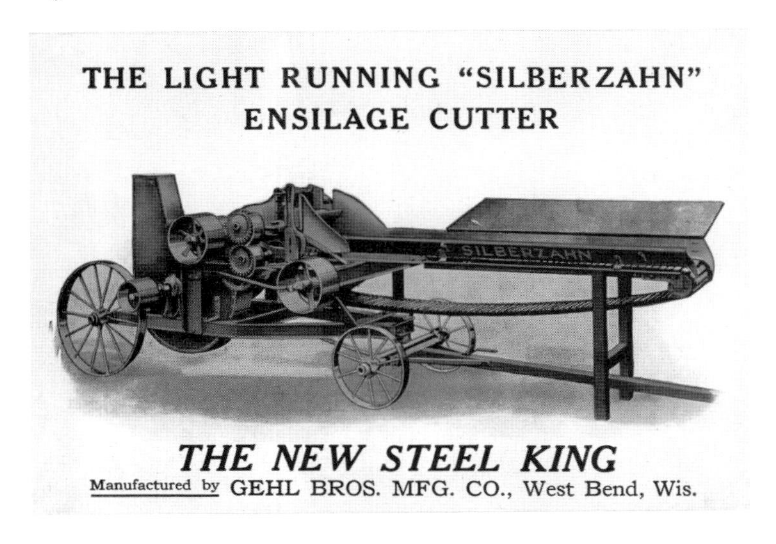

Farmers found instant satisfaction in the Company's steel-frame cutters. The steel frame did not depreciate nearly as quickly as a wood frame.

John W. Gehl's partners in the new venture were Henry J. Thoma and Peter Berres. Thoma was John W. Gehl's wife's younger brother; Berres was a well-known local contractor and manufacturer.[48] As part of the purchase, Berres agreed to move his wagon works to the Silberzahn factory.[49]

On March 16, 1902, John Gehl, Berres, and Thoma met in the offices of the Silberzahn Manufacturing Company in West Bend with the Silberzahns to negotiate the transfer of stock. Charles Silberzahn sold Gehl, Berres, and Thoma eighty shares of his stock, which were split evenly among the three. Homer and Louis Silberzahn quickly resigned, and the new owners took control that afternoon. Charles Silberzahn soon followed, resigning as president in August 1902 after relinquishing his remaining shares in the Company to Gehl, Berres, and Thoma.

Despite the departure of Charles and his two sons in 1902, the Silberzahn family remained shareholders in the Company for several years through shares held by the estate of Charles' deceased son, Charles A. Silberzahn, in a trust for his wife, Amelia, and two minor children. The patriarch, Charles Silberzahn, lived nearly another twenty years, dying in West Bend in the spring of 1921, just short of celebrating his ninety-third birthday.[50]

Three of the four original Gehl brothers (from left to right): Henry, John, and Mike. The fourth brother, Nick, stayed with the Company for only a short while before moving out-of-state.

The Berres-Gehl Manufacturing Company

John W. Gehl and his partners made few operational changes in the Company in their first years of ownership. They did increase the capital stock by $25,000 to issue new stock to pay off Charles A. Silberzahn's widow and to allow for more family ownership. By 1904, John Gehl and Peter Berres owned forty-seven and forty-six shares, respectively, of the renamed Berres-Gehl Manufacturing Company. John's brother, Nicolaus, then came into the Company as a shareholder, replacing Henry Thoma.[51] John and Nicolaus Gehl were joined in ownership of the Company in 1904 and 1905 by brothers Michael and Henry Gehl, who bought out Berres' stock ownership.

The focus on ensilage and feed cutters remained constant throughout the ownership transitions that occurred just after the turn of the century. The first Gehl ensilage cutter was built and sold in 1903.

In late April 1906, the four brothers renamed the business after themselves.[52] The Gehl Brothers Manufacturing Company was literally built on the ashes of the former business.

The 1906 Fire and After

In April 1906, the Gehl Brothers Manufacturing Company plant in West Bend was destroyed by fire. The blaze consumed the factory, foundry, and all inventory. The insurance on the plant wasn't enough to rebuild the plant and restart the business, so the four Gehl brothers pooled all their savings, borrowed where they could, and sold more stock to finance the rebuilding. "There wasn't fifty cents between the four of us," Mike Gehl said years later. "We sold some stock around town, borrowed from the banks, and finally succeeded in building the factory up again."[53]

The Gehl brothers decided to rebuild at the same location on Water Street. The West Bend Lumber Company stood between Gehl Brothers Manufacturing Company and the Northwestern Union Railway tracks, which radiated out from what later became the Chicago & Northwestern Railway depot immediately to the north of the plant. The Company would take advantage of the proximity of the railroad years later by constructing a railroad dock on the inside of the building that could hold as many as five or six railroad boxcars at a time.[54]

The sale of stock brought a sense of community ownership to Gehl Brothers Manufacturing Company. The four Gehl brothers retained ownership of 175 of the 241 shares outstanding. Many of the outside shareholders were local businessmen, including Andrew Pick, Stephen Mayer, and B. C. Ziegler, who owned five shares of stock each.[55] Ziegler, Pick, and Mayer would make West Bend the capital of the nation's aluminum cookware industry during the first half of the twentieth century.[56]

The renamed company's fortunes began to turn around in 1908, when Gehl Brothers Manufacturing Company introduced a

The aftermath of the 1906 fire. Despite severe financial difficulty at the time, the Gehl brothers were able to recover with the help of personal financing and the sale of stock.

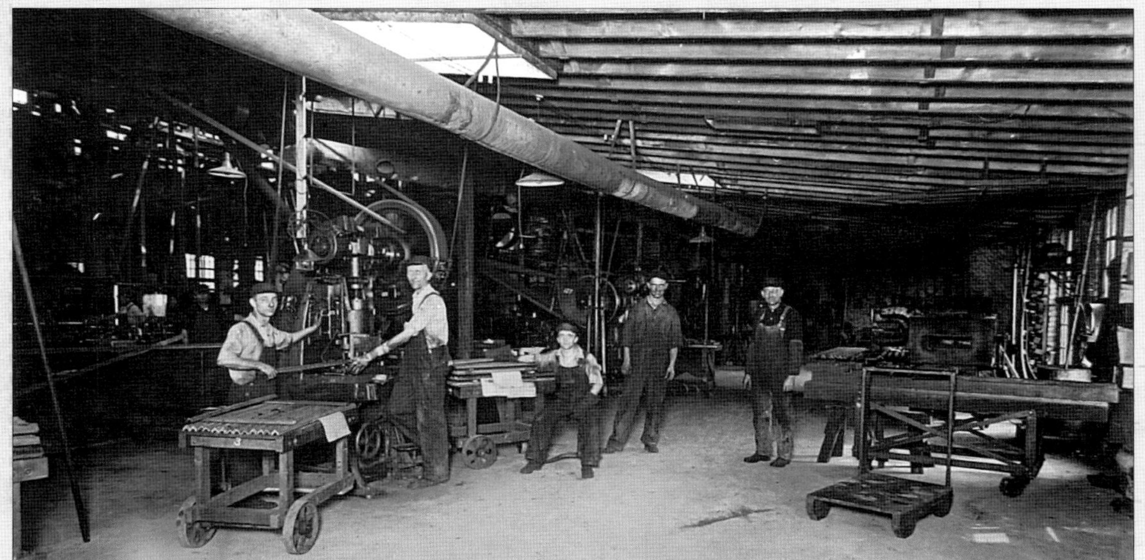

Top: Gehl Brothers Manufacturing Company, as it appeared after being rebuilt following the 1906 fire.

Middle: The 1914 Gehl manufacturing crew, before plants were mechanized and when machines were made by hand.

Bottom: The railroad arrived in West Bend in 1873.

Square steel-enclosed elevators were introduced into the Gehl product line in 1908. The carrier provided a sturdy and reliable means of delivering feed to silos.

Silberzahn cutters were typically shipped to dealers aboard flatbed railroad cars.

wood-stave silo into its product offering. Although the Company only assembled the silos for farmers in the immediate area, they advertised their growing line of mechanized ensilage cutters and carriers as the perfect machinery for filling silos on the nation's dairy farms.

Wisconsin, at the turn of the twentieth century, was the nation's poster child for silo construction. The state had more silos than any other state in the union, and a silo alongside a barn was quickly becoming the signature of the dairy farm nationwide. Silos provided a sealed, efficient method of storing fodder for a dairy herd, allowing dairy farmers to keep their cows milking year-round.

The Wisconsin Department of Agriculture and the Agricultural Extension Agency at the University of Wisconsin worked with farmers across the state to design and build silos, which were typically made of wood, brick masonry, or concrete. In 1908, the Wisconsin State Fair erected a model dairy barn at the Fairgrounds in Madison and included two silos, one built of wood and the other built of brick. Thousands of farmers passed by the display, and by 1916, there were 60,000 silos in the state.[57]

The four Gehl brothers, having grown up around dairy farming in Washington County, realized that silos would become a ubiquitous part of the landscape in Wisconsin and other dairy-producing states. As a result, between 1908 and 1912, they introduced a number of implements designed to speed up the silage process and help farmers fill their silos quickly and efficiently.

In the years following the rebuilding of its West Bend factory, Gehl Brothers Manufacturing Company introduced a larger feed cutter and an elevator. By 1912, the Company had introduced an advanced-design, engine-powered re-cutter for malt grain, corn cobs, and stalks, and also a new silo filler, which quickly developed into the standard for the farm equipment industry.

In 1914, the Company sold six models of ensilage cutters, which could be hooked up to gasoline- or steam-powered engines of up to twenty horsepower. The larger models, when equipped with an enclosed steel carrier, could fill silos up to thirty-six feet tall.[58]

Gehl Brothers Manufacturing Company built this cutter to operate more efficiently and equipped it with a blower to elevate feed into the silo.

{ THE LUCKY THIRTEEN }

Americans were introduced to farm mechanization in the years just prior to World War I. Cheap, reliable gasoline piston engines made possible a new source of motive power for the agricultural implement business. Machinery such as the ensilage cutter and the accompanying carrier could be connected to a gasoline engine, replacing several men and horses. Even more important, tracked or high-wheel tractors could pull implements across muddy ground and allow farmers to till, plant, weed, and harvest their fields without having to rely on teams of horses, mules, or oxen.

This original Gehl tractor was an experimental model being tested on the Kissinger farm near Jackson, Wisconsin, in approximately 1916. Harry Kissinger, pictured sitting on the tractor seat, was an acquaintance of Henry and Mike Gehl and often tested machinery for the Company during that time. *(Courtesy of Marvin Kissinger)*

Wisconsin proved to be an incubator for the farm tractor industry in the 1890s and early 1900s. J. I. Case pioneered a gas tractor engine at its Racine plant in 1894. Three years later, Charles W. Hart and Charles H. Parr, who had recently graduated from the University of Wisconsin, started the Hart-Parr Gasoline Engine Company in Madison. The two soon moved the factory to Hart's hometown of Charles City, Iowa, and made their first tractor in 1901.[65] Allis Chalmers, which had been formed by the merger of two Milwaukee-area companies in 1901, built its first tractor at its West Allis plant in 1914.[66]

Henry Ford, who had put the nation on wheels in 1913 when he began producing the first Model T on the Ford Motor Company's moving assembly line at Dearborn, Michigan, began making Fordson Tractors in 1917. John Deere began manufacturing tractors when it acquired the Waterloo Gasoline Engine Company in Iowa in 1918.[67]

Like many implement manufacturers of the day, Gehl Brothers Manufacturing Company experimented with producing a gasoline-powered tractor in the years just before World War I. Tractor fever was raging across Wisconsin and nearby Iowa in 1917, and just about every machinery manufacturer of the day was producing a tractor.

Gehl had just come off the successful 1917 introduction of the Gehl silo filler, which featured a larger feed cutter and a blower to blow the chopped material into the silo. Henry Gehl, who spearheaded the firm's research and development efforts at the time, turned his attention to the design of a tractor.[68] Gehl Brothers Manufacturing Company secured a license on a gasoline engine manufactured in Wisconsin and introduced a tank-type machine with large spikes studding the steel wheel for rear drive. The operator perched precariously on a saddle-type steel seat located behind the mammoth rear wheel.

About the time that Gehl Brothers Manufacturing Company introduced its new tractor, President Woodrow Wilson asked Congress to declare war on Imperial Germany in April 1917. The resulting shift to a wartime economy meant that the West Bend firm could not get engines or steel to build the tractors. The line was quietly dropped, and Gehl concentrated on the production of forage-harvesting equipment.

Years later, Richard Gehl offered the opinion that Gehl Brothers Manufacturing Company had possibly been saved from a serious mistake by Congress' declaration of war. Then the firm's vice president of marketing, Richard noted that it was "probably a stroke of luck for the Company that Uncle Henry was not too successful and, more important, that only about a dozen of these machines were produced."[69]

Company legend has it that only slightly more than a dozen of the tractors were produced, which led subsequent generations to dub the machine the "lucky thirteen."

The Gehl tractor on display in a parade through downtown West Bend.

The Company recommended that "in filling the silo, the corn should be cut into about one-half-inch lengths, and should be cut cleanly and evenly. Corn put in the silo half-shredded will not pack tightly and so will not exclude the air, which will cause it to rot. When the corn is cut fine, and evenly, it packs together a great deal better, making better ensilage, besides increasing the capacity of the silo. When well-packed, the air is entirely excluded and the corn by a process of fermentation develops into a very palatable and easily digested food, upon which animals thrive very well."[59]

Gehl Brothers Manufacturing Company obviously thought that its mechanized feed cutters and carriers were far superior to others on the market. Dairy farmers in Wisconsin and the Midwest thought so, too. Although no sales figures survive from the years prior to World War I, a 1914 photograph gives an indication of the scope of sales at the time. The photograph shows Gehl cutters piled on pallets in the siding yard at the plant in West Bend, awaiting loading onto rail cars.[60]

The local businessmen in West Bend played a large role in the survival and expansion of the industry within the community. The two men in front are John W. Gehl (left) and B. C. Ziegler. *(Reproduction from the collection of the Washington County Historical Society, #016,720)*

The World War I Years

Gehl Brothers Manufacturing Company, like much of the nation's farm sector, prospered during the farm boom of the World War I years. Corn, wheat, and other agricultural commodities hit record highs in 1917 and 1918, and dairy farmers found a ready market for their milk and cheese at the hundreds of military installations that dotted the countryside. The creation of a federal Agricultural Extension Service in 1914 and the inroads made by proponents of scientific farming in the early twentieth century increased yields and farm efficiency during the war years.[61]

By thoroughly studying the ensilage cutting process, including its strengths and weaknesses, the Company was able to ensure a better machine and greater value for the customer (c. 1917).

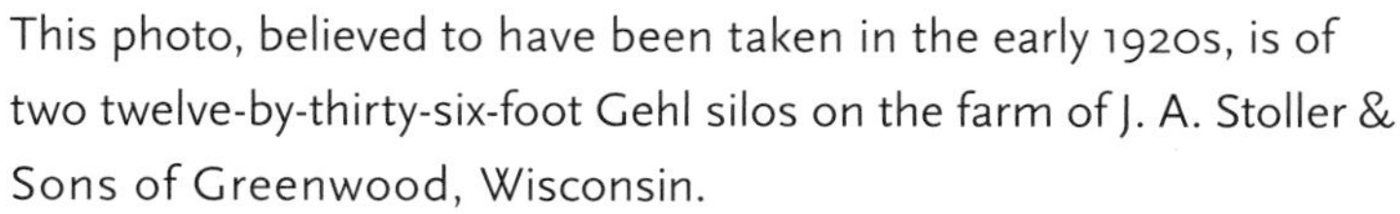

This photo, believed to have been taken in the early 1920s, is of two twelve-by-thirty-six-foot Gehl silos on the farm of J. A. Stoller & Sons of Greenwood, Wisconsin.

A Gehl No. 17 ensilage cutter filling its 117th silo on the Jaeger Farm in Campbellsport, Wisconsin.

When the war began in Europe in 1914, Gehl Brothers Manufacturing Company advertised itself as "the largest exclusive ensilage cutter factory in the world."[62] The Company's plant sprawled across several square blocks of West Bend and included a main factory, a foundry, five product warehouses, and several parts and lumber warehouses. The Company employed one hundred workers year-round.

The Gehl Brothers Manufacturing Company office as it appeared in 1919.

The prosperity of the war years was reflected in the Company's balance sheet. In 1912, Gehl Brothers Manufacturing Company's total capital and surplus had reached $119,000, the first time it had exceeded $100,000. By 1915, total capital and surplus had grown to $150,000. Three years later, in 1918, capital and surplus was just under $250,000.[63] Capital and surplus, which included common and preferred stock values, inched upward again in 1919 and peaked at just over $265,000 in 1920.[64]

But with the onset of the farm depression in 1921, followed by the Great Depression in 1929, Gehl Brothers Manufacturing Company wouldn't again reach those annual capital and surplus figures for another twenty years.

GEHL
ENSILAGE CUTTERS

Chapter Three

GROWTH AND *Contraction*

1920–1940

The 1920s were terrible years for the American farm economy.

The boom that had characterized the 1910s and the World War I years evaporated almost as soon as the Peace Treaty of Versailles was signed, ending the war. Historians dubbed the decade following the end of the First World War the "Roaring Twenties." But for the nation's farmers, the decade was an almost endless source of bad news.

The war had been a boon for America's farmers. Agricultural production in Europe and Russia started to decline in 1914 and accelerated downward from there. American farmers planted "fence row to fence row," and that meant that much more marginal land—such as the cutover regions of northern Wisconsin and adjacent Minnesota—was brought into agricultural production. Farmers increasingly went into debt to purchase land and equipment, which rippled back to the bottom line of implement manufacturers such as Gehl Brothers Manufacturing Company.

European agricultural production, however, recovered much more rapidly than anyone had expected, and the results were disastrous for American farmers. The U.S. farm product price index fell by more than half from 1920 to 1921, and real average net income per farm dropped by nearly 75 percent in the same period of time.[70] Farm income recovered to near 1920 levels in 1925, but then began a second dive that continued through the end of the decade and put farm incomes at historic lows at the beginning of the Great Depression.

Farm mortgage foreclosures began an almost unbroken fifteen-year rise in 1920, and farm population and employment dropped throughout the decade. The drop in farm employment during the decade was influenced by the increasing mechanization of agricultural production. In 1921, the lowest point of the postwar farm depression, Gehl Brothers Manufacturing Company introduced a disc-type silo filler with a flywheel cutter, which could be powered by either a gasoline engine or a tractor equipped with a belt pulley.

The drop in farm population and employment was somewhat offset by the gain in productivity. Wisconsin, which had surpassed New York as the top dairy state around 1915, benefited from advancements in transportation and refrigeration. Refrigerated rail tank cars made it possible for Wisconsin dairy farmers to ship their milk all the way to plants in the Mid-Atlantic States.

It's "delivery day" at Sliter & Rivers on August 29, 1924.

An early 1920s display of the Company's pride in its flagship product.

Advances in pasteurization and home refrigeration created an urban market for milk that had never existed before. Meanwhile, average annual yield per cow continued to climb from 3,050 pounds of milk in 1890 to 4,500 pounds in 1950.[71]

A measure of the growing popularity of dairy farming in the counties north and west of Milwaukee came one early June day in 1923, when more than 10,000 visitors flocked to the Schroeder Farm near West Bend to attend the second annual Wisconsin Dairymen's Field Day. The event, organized by John W. Gehl, B. C. Ziegler, Paul Bast of Germantown, and L. P. Rosenheimer of Kewaskum, featured cold Wisconsin milk and buttermilk and ham-and-cheese sandwiches. More than 1,300 automobiles were parked on the Schroeder Farm grounds, and visitors toured the barns of Cedar Lake Farm to inspect Schroeder's herd of registered Holstein cattle.[72]

The Financial Picture

By the 1920s, Gehl Brothers Manufacturing Company was one of the leaders in the region for farm implement manufacturing. But early in the decade, the Company was losing money.

A measure of the severity of the 1920–1921 depression in the farm economy came at the end of 1921. In November 1921, Gehl Brothers Manufacturing Company reported a loss of $60,000.[73] The next year, the loss was $70,000, and in 1923, the loss was $53,000. In the decade before 1921, the Company had reported an average annual surplus near $25,000.[74]

The Company had financed its growth during the previous decade through capital stock authorizations. In October 1913, Gehl Brothers Manufacturing Company declared a stock dividend of $40,000 and had authorized a bond issue of a like amount the previous May. The Company's capital was increased to $250,000 in 1917, and the issuance of an additional $80,000

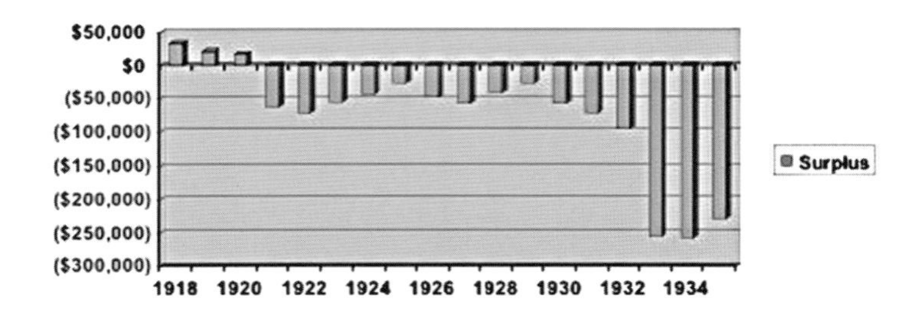

Surplus of Gehl Brothers Manufacturing Company at the close of the fiscal year for 1918–1935.

The 1921 Gehl model B4 silo filler shown in the Gehl plant.

Despite continuous improvements to the design of the Gehl ensilage cutter, certain patented features of the earliest models remained constant. The patented enclosed-steel carrier was long offered as optional equipment, and the patented reverse gear appeared exclusively on Gehl ensilage cutters.

The original Gehl hammer mills were built in three styles: the plain mill required all roughage to be fed through the same opening; the combined ear corn crusher and feed mixer (as shown) crushed ear corn separately and then mixed it with the grain; and the combination mill handled grain and ear corn separately or all together.

in bonds was authorized in 1920. Capital was increased again in December 1921 to $350,000, to offset some of the decrease in surplus in 1920 and 1921, and in 1923, Gehl Brothers Manufacturing Company issued another $80,000 in bonds.[75]

Gehl Brothers Manufacturing Company had a reputation for rewarding those who had invested in the Company's rebirth following the 1906 fire. Between 1909 and 1921, the firm paid $152 in dividends on each share of common stock. The regular dividends, averaging a little over $10 a year per common share, meant that there was no lack of local investors interested in purchasing new issues of common stock.[76]

The Gehl Hammer Mill

The growth of the dairy industry during the 1920s created a demand among farmers for an easier method of grinding homegrown grains. The Company responded in 1929 with the development and introduction of the Gehl hammer mill.[77]

Corn had long been the silage of choice for dairy farmers, but it wasn't until the early twentieth century that the nation's agricultural colleges had discovered that livestock had a difficult time digesting corn. Cattle, particularly, with their four stomachs, did best on coarse diets and high fiber. The solution was the hammer mill, which allowed the farmer to crack the corn kernels into small pieces, which were easier for cows to digest.[78]

The stationary model of the Gehl hammer mill found immediate success on dairy farms in the Midwest and Northeast. The Company supplemented its stationary hammer mill with a portable truck-mounted hammer mill that became an icon in dairy farming regions. A common sight in those years, as explained in a 1984 history of the Company, "was the custom operator with a Chevrolet truck, axle deep in mud or snow, going from farm to farm grinding feed."[79]

Manure handling provided further opportunities for Gehl Brothers Manufacturing Company in the 1920s and 1930s. Beginning in 1927, and continuing until after the end of World

The switch from stationary to portable hammer mills created the opportunity for individuals to start their own businesses by traveling between farms grinding feed.

War II, the Company manufactured manure spreaders. The Gehl manure spreader came with auto-style steering instead of the common wagon-style steering that had characterized older models. Auto-style steering allowed the farmer to position the spreader more precisely beneath a manure carrier or in a small barnyard. The auto-style steering also enabled the operator to maneuver the manure spreader along narrow rural lanes and through farm gates.[80]

Gehl Brothers Manufacturing Company's introduction of a line of manure spreaders that could be pulled by a farm tractor replaced the old general-purpose wagon with its farmhands scooping manure from the bed onto the fields behind. Gehl employees joked that although the Company stood behind all of its equipment, nobody would stand behind its manure spreader.

In 1932, Gehl Brothers Manufacturing Company introduced one more product line, its first piece of equipment that was not intended for sale exclusively to the farm industry. Gehl coal stokers were marketed to both residential and farm consumers, primarily for heating homes.[81] In the 1930s, most homes in Wisconsin and the Upper Midwest were heated by coal-fired furnaces. The coal was mined in the Pennsylvania and West Virginia coal fields, hauled by rail to ports on Lake Erie, and backhauled up the Great Lakes to coal docks in Milwaukee, Chicago, Duluth, and other Lake ports.[82]

The introduction of residential electric power in the 1920s and 1930s allowed homeowners to fill a stoker with coal once or twice a day. A worm-like gear inside the stoker evenly fed coal to the furnace over a period of hours, eliminating the need to shovel coal directly into the furnace several times during the night. Gehl marketed its line of stokers in Milwaukee and other nearby communities until it sold the line following World War II.

Throughout the 1920s and 1930s, the backbone of the Gehl Brothers Manufacturing Company product line was the Gehl ensilage cutter and silo filler. The West Bend firm had ample competition for these product lines. A number of implement manufacturers made cutters and silo fillers in the 1920s and 1930s, including the Advance-Rumely Company in LaPorte, Indiana; J. I. Case in Racine, Wisconsin; McCormick-Deering in Chicago, Illinois; the Smalley Manufacturing Company in Manitowoc, Wisconsin; and the J. S. Rowell Manufacturing Company in Hartford, Wisconsin.

The hand-firing of furnaces was eliminated with the development of the coal stoker. Gehl stokers tended the fire in homes both day and night and produced significantly less smoke, soot, and fumes than hand-firing.

The Gehl manure spreader evenly distributed fertilizer while saving farmers valuable time in comparison to traditional methods of fertilizing.

The Gehl No. 27 model boasted a seventy-bushel capacity, considerably larger than the other Gehl model of the time, the No. 26, which provided a sixty-bushel capacity.

The Gehl disc-type silo filler's low speed meant less vibration, longer life, and lower power requirements.

A fully-restored Gehl No. 27 manure spreader.

STOKERS
DOMESTIC
COMMERCIAL
INDUSTRIAL
COAL-CONVEYORS
AUTOMATIC CONTROLLS

Phone WEst 3-7748

GEHL STOKERS CO.
3521-3523 WEST VLIET STREET
MILWAUKEE 8, WISCONSIN

OIL BURNERS
DOMESTIC
COMMERCIAL
INDUSTRIAL
BOILERS
FURNACES

SOLD TO Cash Sale
DATE 3-15-56
SHIPPED VIA
YOUR ORDER NO.
TERMS
DATE SOLD 3-15-56

QUANTITY	DESCRIPTION	PRICE	AMOUNT
2	Belts @ #1V80	$1.24	$2.48

Paid 3-15-56 J Berner

A 1956 receipt for repair parts on a Gehl stoker, purchased through the Company's Milwaukee-area stoker distributor. The Gehl stoker line continued well into the 1950s.

McCormick-Deering and its wholly owned subsidiary, International Harvester, was the largest competitor. A vertically integrated giant that owned steel mills in Wisconsin and Illinois, iron mines in Minnesota, coal and coke works in Kentucky, and manufacturing plants across the Midwest from Ohio to Minnesota, McCormick-Deering dominated most of the agricultural implement markets it entered. In 1925, the Chicago-based company reported a net profit of $19 million and $100 million in common stock.[83]

In 1923 and 1924, the Engineering Department at the University of Wisconsin in Madison worked with Gehl Brothers Manufacturing Company and five other U.S. manufacturers, including McCormick-Deering, to test ensilage cutters and silo fillers. The tests revealed that the Gehl flywheel and cylinder cutters compared well with all of the other makes and models in terms of cutting efficiency and tons-per-hour production. Through the tests, which were reported in *Agricultural Engineering Magazine,* Gehl Brothers Manufacturing Company learned of ways to improve the efficiency of its flagship line of agricultural machinery.[84]

Transitions

Management at Gehl Brothers Manufacturing Company had remained in the hands of the sons of Michael Gehl since just after the turn of the century. Nicolaus Gehl had sold his shares of Company stock to his brothers in 1919, resigned from the Board in 1924, and headed west. Nick, as he was known in the family, died in Denver, Colorado, in 1928 at the age of fifty-nine.

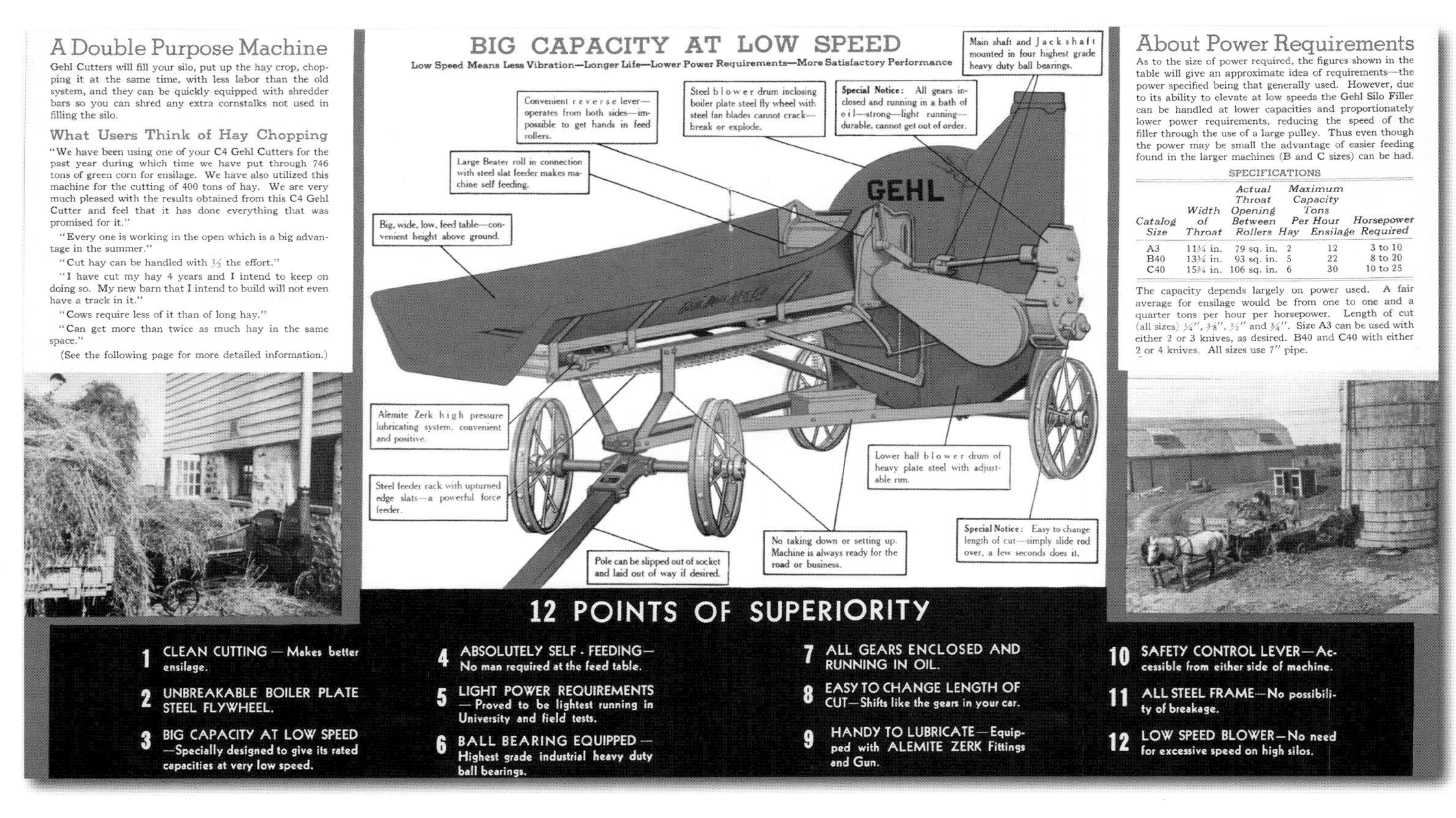

The design of the Gehl cutter improved over the years through the Company's experience in the ensilage-cutting business and through the needs expressed by its users.

The Gehl size "B" silo filler, in use on the Preskon Farm in Eau Claire, Wisconsin, in 1928. Often, users would run the silo filler with a five-horsepower motor, as shown here with a "three-horsepower wagon," which allowed the machine to run at a satisfactory capacity with light power and at a low speed to keep it from "choking."

Brothers John, Henry, and Mike operated the Company for nearly a half-century. John, the oldest of the remaining brothers, was the general manager. Henry, the youngest, handled engineering and research and development projects. Mike was in charge of sales and marketing.[85] The three were assisted in the management of the Company by Arthur Merriam, the longtime office and advertising manager.

After World War II, Henry and John ran the Company before turning it over to the next generation. John Gehl died in West Bend in 1951 at the age of seventy-nine. After he retired from active management of the firm, Mike Gehl served as mayor of West Bend from 1948 to 1954.[86]

By the early 1930s, another generation of Gehls was arriving at the family company on Water Street. Four of John W. Gehl's sons and Henry Gehl's son all joined Gehl Brothers Manufacturing Company between 1930 and 1940.

The sons of John and Henry Gehl were born between 1901 and 1914 and grew up in West Bend. The sons of John Gehl attended Holy Angels Roman Catholic School, which was operated by the School Sisters of Notre Dame.[87] Albert, the only surviving member of the second generation still alive as this sesquicentennial history of Gehl Company went to press, is the second youngest of the children of John W. and Mary Gehl. Richard, Mark, Albert, and Carl all joined Gehl Brothers Manufacturing Company. Brother Frank was a mortician. He went to a school in Chicago and worked for a mortician in Boston and later operated the Kapfer-Gehl Funeral Home in West Bend, which was also a furniture store. He semi-retired to Florida and operated a small store in the Ft. Lauderdale area in his later years.[88]

John W. and Mary Gehl's only daughter, Anita, "worked at a tuberculosis sanitarium in West Allis. She went back to

{ THE FIRST FORAGE HARVESTER }

The U.S. recovery from the Depression helped Gehl Brothers Manufacturing Company recuperate from the economic doldrums that had plagued American manufacturers for nearly a decade. When recovery began to take hold during the late 1930s, there was a significant expansion of the dairy-farming sector of the agricultural economy, which continued into 1940 and 1941. The Japanese attack on the U.S. Naval Base at Pearl Harbor on December 7, 1941, and the American declaration of war the next day, created a farm labor shortage by the spring and summer of 1942.

To free up as many farmhands for service in the military during the war, the United States needed a labor-saving method of putting up hay and corn silage.[111] After a quarter-century of manufacturing silo fillers, Gehl Brothers Manufacturing Company had the research and manufacturing expertise to develop its first field forage harvester in 1942. The first forage harvester was essentially a silo filler placed on wheels and pulled across the fields with a high-sided wagon box trailing behind. With a sixteen-inch throat, hay pickup head, and flywheel-type cutter design, the Gehl forage harvester revolutionized forage harvesting.[112]

Farmers loved the new Gehl forage harvesters. Many reported to the Company that it was far easier to unload chopped corn into a forage blower to fill a silo than it was to throw the bundled corn stalks into the silo filler. Making hay became a far less onerous task. Before the introduction of the Gehl forage harvester, a crew of two or three farmhands loaded the hay into a wagon, conveyed it to the barn, and then manually fed the chopper and filled

The first Gehl forage harvester built for production purposes, being operated in 1942 on the Gehl homestead farm in the town of Hartford near St. Lawrence. The forage harvester is being operated by Joseph Gehl and the horse-drawn wagon by his wife, Eleanor Gehl. Due to an abundance of problems with the initial 236 machines built in 1942 and 1943, those machines were brought back to the factory, completely rebuilt, and returned to their owners. Although it was an expensive move, it provided great satisfaction among Gehl customers.

The main unit of the forage harvester connected to a gas engine.

the silo.[113] A 1984 history of the Company noted that "it was downright luxurious for one man to be able to pull a Gehl chopper and wagon into a hay field and [chop and] blow two acres of hay into the wagon."[114]

Gehl Brothers Manufacturing Company began production of the new line of forage harvesters in 1943. Late in the year, the Company introduced a single-row corn head attachment for the forage harvester.[115]

As the years went on, Gehl Brothers Manufacturing Company made significant improvements to the Company's line of forage harvesters. In 1955, the Company introduced a front-unloading forage box that delivered chopped forage directly to a hopper forage blower. Throughout the 1950s and 1960s, Gehl Brothers increased the capacity of the blowers on its forage harvesters, offering two-row corn heads and higher capacity hay pickup heads, and even introducing a self-propelled model.[116]

The 1968 introduction of the cylinder-type "cut-and-throw" model CT300 forage harvester made Gehl Brothers Manufacturing Company competitive with companies such as Fox and New Holland. Terry Lefever, who retired in 2007 as the Company's special events manager, sold the cut-and-throw forage harvesters in Montana and Wisconsin during the late 1960s and early 1970s. He recalled that the new forage harvesters took Gehl Brothers from a 7-percent market share in the mid-1960s to a 25-percent market share in the mid-1970s.[117]

The 1973 introduction of the combination cylinder-type "cut-and-blow" model CB400 forage harvester with spinner and blower was a further refinement of the Company's leadership in the design and development of forage harvesters.

During the 1970s, Gehl Brothers Manufacturing Company introduced other innovations, including a push-type self-propelled forage harvester that mounted on a tractor's front end and was steered from a control panel in the cab.[118] Other innovations during the decade included a three-row corn head, a snapper head attachment, and new built-in sharpeners for the cutting cylinder knives.[119]

The Gehl forage harvester eliminated nearly all of the manual labor previously required to make hay and silage.

The first and second generations of Gehl brothers. From left to right are: Dick, John W., Mark, Mike, Al, Henry, and Carl.

school to become a dietitian and got a job at the Milwaukee County General Hospital. She lived at Twenty-fourth and Wells in Milwaukee across from the Elks Club and ended up living at the Milwaukee Catholic Home," Albert recalled.[89] A brother, Alfred, born in 1904, died in infancy the following year.[90]

Richard ("Dick"), the oldest, was the first to join the Company. Following his graduation from Campion Preparatory School in 1918, he initially did not want to join the family business, and his father, John Gehl, arranged for Dick to work as a teller at the Western Bank in Milwaukee, the Gehl Brothers Manufacturing Company's bank. Dick eventually joined the family firm as a salesman in 1921, and by the 1930s, Dick Gehl was sales manager of the Stoker Division.[91] Mark graduated from Marquette University with a degree in engineering and joined Gehl Brothers Manufacturing Company in the mid-1920s. Albert graduated from Marquette with a degree in accounting in 1931 and immediately joined the family firm. "Marquette had an enrollment of 3,000 students," Albert said. "It always amazed me that the school had a larger population than West Bend did at the time."[92] Carl, the youngest of John and Mary Gehl's children, earned both his undergraduate and law degrees from Marquette and worked at Hardware Mutual Insurance Company (now known as Sentry Insurance) before joining Gehl Brothers Manufacturing Company in the late 1930s.

Albert and Mark began to take on ever-larger shares of management responsibility during the 1930s, and by the early 1940s, the second generation of four Gehl brothers were at the helm of the family business. Mark Gehl was plant manager and in charge of engineering and manufacturing, assisted by his cousin Bernard ("Ben") Gehl. Ben, Henry's son, had earned his engineering degree at Marquette and joined the Company in the early 1930s, about the same time as his cousin Albert. Albert was the Company's accountant, handling the Company's business and financial affairs, and was later elected president by his brothers and uncles. Carl Gehl's responsibilities included development

Just prior to the second generation of Gehl brothers joining the Company, the Gehl Champion was introduced around 1923. The Champion received its name because the Company maintained that its cutter was the "champion" of all cutters.

of the export markets and oversight of personnel, labor relations, and legal affairs. Dick Gehl assumed the responsibility for marketing and selling the factory's output.[93]

By the time the second generation of Gehl brothers took over management of the Company, it had undergone wrenching change brought about by the Great Depression.

Surviving the Great Depression

Gehl Brothers Manufacturing Company survived the Great Depression, but not without struggle and effort. The election of President Franklin Delano Roosevelt in 1932 found the country in the throes of a credit crisis unparalleled in American history. Roosevelt's first one hundred days in office in the spring of 1933 were characterized by bank failures, massive unemployment, and foreclosures from one coast to the other. Americans lost confidence in the nation's banking system. Wisconsin farm income dropped from $350 million in 1930 to less than $200 million two years later, propelled downward by agricultural commodity prices that plunged by half during the same period.[94] Wisconsin was not immune to the drought that plagued the Midwest during the first half of the 1930s; the state suffered the worst drought in its history in 1934.[95]

In the early 1930s, farmers continued to order and purchase new Gehl hammer mills and Gehl manure spreaders. But by 1933, milk strikes in the Fox River Valley and the growing farm credit crisis in the Upper Midwest heralded difficult times ahead for Wisconsin's agricultural implement manufacturers.

Brothers Lloyd and Don Theisen grew up during the Depression in the western North Dakota town of Sentinel Butte. "Dad was a jack-of-all-trades," Lloyd Theisen said. "He even was the town butcher. We ate pretty well, but we both thought the Depression was a part of life."[96]

Lloyd graduated from high school in 1934, when thousands of windburned North Dakota farmers stormed the state's new Capitol Building at Bismarck to reinstall the ousted Bill Langer as governor and to demand relief from the economic woes that were dragging down the state.[97] That summer, the dust rolled across the Northern Plains like a giant curtain, and triple-digit temperatures for weeks on end burned crops in the field.

Taken around 1921, the same year Dick Gehl joined the Company as a salesman, this photo shows the Gehl salesmen in front of their cars. These salesmen were relied on to travel the country, building the Company's reputation beyond the community in which it was located.

From left to right are: Mike Gehl, Ed Bauer, Chas Toubier, August Wegner, Lawrence Huber, Al Ciriacks, Charley Chapps, George Hetzel, Shorty Hoffman, Otto Weber, Art Merriam, Walter Barth, Christ Klapper, John W. Gehl, Walter Rilling, Peter Bauer, Barney Ciriacks, Charley Mauer, Ed Kuester, John Steilen, Al Ruff, Curley Grevy, Oppie Appenzeller, John Degen, A. Kieckhafer, Walter Oller, Walter Kahl, Clarence Hahn, and Henry Gehl.

The Gehl No. 26 model, the lower-capacity manure spreader of the time, sitting between buildings in the Gehl yard.

Lloyd Theisen rode a potato truck from Beach, North Dakota, to Moorhead, Minnesota, and then hopped a freight train from Moorhead to LaCrosse, Wisconsin. He rode a coal car on a passenger train from LaCrosse to Milwaukee. He had an uncle in Germantown, who put Lloyd to work for one dollar per day and room and board, and the North Dakotan was happy to get it.[98]

The Germantown uncle got his nephew a job at the West Bend Aluminum Company in 1936, but three months later, he was laid off. In March 1937, Theisen recalled, "My Uncle Frank got me a job at Gehl Brothers, where I worked as a molder in the foundry. The machine shop was in the front of the West plant, and the foundry was in the back."[99]

Don Theisen joined his brother at Gehl Brothers Manufacturing Company later in 1937, only to be laid off for several months in the spring of 1938. When he was called back, he did what needed to be done, no questions asked.[100]

"We had been making hammer mills," Don Theisen recalled. "We also made a silo filler for Sears Roebuck. It was called 'The Reliable.' The only difference was The Reliable had [a] one-and-a-half-inch angle iron on the frame versus two-inch angle iron on the Gehl silo filler. We also made a manure spreader and a ten-inch stationary hammer mill. We were also making the stokers in the late 1930s. I worked as a spray painter on the stoker line in 1938 and 1939. They baked the enamel with heat lights."[101]

Despite the best efforts of the Roosevelt New Deal, the nation's economy continued in the doldrums. By the fall and early winter of 1933, nearly a quarter of the American work-force was unemployed. In a land of supposed plenty, people were going hungry; soup kitchens sprung up in American cities, including Milwaukee.

The situation was not as desperate in West Bend, partly because the city had a diversified employment base, including the West Bend Aluminum Company, Amity Leather, and Gehl

Left: Three of the most popular Gehl products of the time are represented in this photo: the feed cutter in the foreground; a Gehl stoker advertisement in the background; and a hammer mill on the left in the distance.

Below: The Gehl brothers continuously made every effort to maintain a feeling of equality amongst all employees of the Company, which included their being present in all areas of the building. Here, Mike Gehl is shown assisting a worker in the foundry with pouring a mold.

Brothers Manufacturing Company. Still, in 1933, the Bank of West Bend—one of three banks in the city—failed, and the teachers in the local schools took a 10-percent pay cut.[102] And for a time in the spring of 1934, it appeared that Gehl Brothers Manufacturing Company might not make it to 1935.

Reorganization

For Gehl Brothers Manufacturing Company, the Depression was a continual struggle. Money was tight, and even though farmers loved the Company's signature line of silo fillers, they rarely had the money to buy the machinery, which meant that dealers were unable to finance large inventories. As a result, Gehl Brothers Manufacturing Company's sales suffered throughout the early 1930s.

Matters came to a head for Gehl Brothers Manufacturing Company in 1933. Agricultural implement sales were down sharply nationwide, and when Gehl stockholders met in West Bend in December 1933 for the Company's annual meeting, Acting Secretary A. L. Merriam "read over the figures for the profit and loss for the year and commented on the great shrinkage in the sales as compared with normal."[103] Even though "overhead expenses had been held down as closely as possible," Mike and Henry Gehl noted that the prospects for 1934 looked little better than 1933.[104]

In an attempt to raise money, the Company issued 2,500 shares of no-par-value stock, which could be sold at any price that the Board "may deem necessary or expedient for the best interest of the corporation."[105] The Company also diversified, operating the Gehl Hardware Store in West Bend, as well as a manufacturer of small concrete mixers and ceramic salt cups for livestock.

By the next spring, Gehl Brothers Manufacturing Company was perilously close to defaulting on a number of its bonds. At a special shareholders meeting on April 25, 1934, the Gehl brothers admitted that "this corporation is unable to pay its existing obligations to its creditors and continue in business without financial assistance."[106] At the special meeting, the Gehl brothers, who controlled about 850 of the Company's 1,520 common shares, presented a plan to the stockholders, bondholders, and creditors that they hoped would avert a bankruptcy.[107] The plan was essentially a trusteeship that would allow Gehl Brothers Manufacturing Company to reorganize and pay off its debt.

Key to the reorganization plan was a $50,000 loan from Bernhard C. Ziegler, West Bend's most influential resident. B. C. Ziegler, the son of a tavernkeeper, had taken over a West Bend insurance agency in 1902 shortly after graduating from high school. Ziegler started making loans to area farmers in 1907 and floated his first bond issue to help build the community's Church of the Holy Angels in 1913. Two years before, Ziegler had joined several associates in forming the West Bend Aluminum Company to make pots, pans, and pie plates. In 1920, he incorporated B. C. Ziegler and Company as an investment banking firm. In 1929, as the Great

The No. 41 Gehl Grind-All hammer mill featured a grinding chamber with a twenty-four-inch diameter and forty-two swinging hammers, providing large capacity at a low speed.

The Gehl Cup Company sold a line of ceramic salt cups used to hold salt as a lick for cattle held in stanchions.

Gehl Brothers Manufacturing Company often manufactured products on contract for other local companies. One product believed to have been manufactured by Gehl and private-labeled was the Johnson tractor heater, pictured here.

B. C. Ziegler

Depression approached, Ziegler sold off all of his stock holdings and paid off his company's debts. When Wall Street collapsed a few months later, Ziegler was sitting on a pile of cash.[108]

B. C. Ziegler was a strong promoter of West Bend. He was also the brother of Henry Gehl's wife, Catherine. So when Henry approached Ziegler about a loan to keep Gehl Brothers Manufacturing Company afloat, Ziegler agreed. Although there was the blood tie, it is also likely that Ziegler was concerned about the impact on West Bend should one of the town's leading employers be declared insolvent.

Ziegler extracted a price for the loan. Gehl Brothers Manufacturing Company had to agree to operate under a trusteeship until their debts were repaid. And they had to agree to give Ziegler proxy control of the Company's finances until the Company was back on its feet.

Ziegler asked Henry Arnfield to take over the financial duties at Gehl Brothers Manufacturing Company, and Arnfield joined the Company as comptroller in 1934. A native of Germany, he had arrived in West Bend in 1920 at the urging of University of Wisconsin classmates Carl Klumb and Walter Malzahn, who persuaded the young immigrant that their hometown of West Bend was a great place to begin a career. Arnfield worked for Carl Pick in an automotive parts business and came to the attention of Ziegler when he worked for the West Bend Company for two years in the 1920s. He then spent the next ten years running his own accounting firm.[109]

Arnfield, who presided over the finances at Gehl for the next twenty-nine years, helped spearhead the remarkable resurgence of the West Bend implement manufacturer in the latter half of the 1930s. The Company had paid off all its debts by 1940, Arnfield said in a 1979 interview, "and has been prosperous ever since."[110]

In 1940, the storm clouds of war were gathering across the globe. It was only a matter of time before the United States would be fully involved.

Chapter Four

WAR AND *Peace*

1941–1960

With the financial problems of the Great Depression behind it, Gehl Brothers Manufacturing Company turned its attention to serving the nation's dairy market, which was itself recovering from the economic upheavals of the 1930s. But before the West Bend implement manufacturer could fully retool its manufacturing and marketing focus, the global conflict that had enmeshed Europe and Asia since the mid-1930s embroiled the United States in the second world war of the twentieth century.

News of the Japanese attack on the U.S. Naval Base at Pearl Harbor hit West Bend on a quiet Sunday afternoon in early December 1941. Overnight, Gehl Brothers Manufacturing Company was enlisted in the nation's war effort.

The war brought major changes to Gehl Brothers Manufacturing Company. The federal government considered the Company a critical war industry asset because of its agricultural implements and their importance to the government's goal of feeding the nation's workers and soldiers. Development of the innovative Gehl forage harvester began in the early 1940s, and the Company was placed on government priority lists for steel throughout the war years.

Albert Gehl recalled that "Gehl Brothers did subcontracting for Milwaukee firms that had contracts with the government. We also made some farm machinery. Farm equipment had a priority for getting steel and other material. We had to get priority from Washington to use a certain amount of steel."[120]

Albert Gehl also recalled that the family-owned company did not subscribe to much of the anti-German hysteria that was sweeping American society at the time. "We had a shipping clerk who was a German immigrant and couldn't get a job in Milwaukee because of his background," Gehl said.[121]

Transitions

Gene Wendelborn recalled that Gehl was already undergoing a transition when he began working at the Company as a teenager in the fall of 1942.

"I was a senior at West Bend High School," Wendelborn explained. "I got the job through the efforts of Henry Gehl. A lot of Company engineers had left for World War II, and he was short of help. The manual arts teacher at West Bend High School, Walter Schuelke, selected two students to apply at Gehl to do engineering work, me and Jerome Hahn, whose dad worked at Gehl. I worked full-time. I would get there in the early morning and work until 10 a.m. Then I left school at 2 p.m. and worked until 6 or 6:30 p.m. plus Saturdays."[122]

Like so many young men of his generation, Wendelborn wanted to serve his country. In November 1942, he approached his employers at Gehl about joining the U.S. Navy. "John Gehl got me a six-month deferment," Wendelborn said, "and the

Several of the most successful Gehl products, such as the Gehl forage harvester shown here with two customers, have their origins in the Company's flagship product, the feeder cutter. And "successful" is the perfect word to describe the Gehl forage harvester. By the 1950s, Gehl Brothers Manufacturing Company was advertising that it was selling more forage harvesters than any other independent manufacturer.

The revolutionary Gehl forerunner forage harvester, as it appeared when first introduced in 1943.

Navy recruiter said that I should finish high school. I graduated in June 1943, and then I decided to go in[to] the Navy. I came back in March 1946 after serving as a quartermaster in the Pacific Theater."[123]

During the more than eleven months early in the war that Wendelborn worked at Gehl Brothers Manufacturing Company, he saw a company that was in the midst of change. "In 1942 when I started, Gehl was transitioning from manufacturing primarily stationary forage handling implements to the mobile type," Wendelborn said. "They had attempted to put together a series of machines, primarily attachments for the forage harvesters, that weren't successful at first. But a number of redesigned attachments were developed by late 1942 and those were successful. We were doing a lot of job-shop work in the Foundry and Sheet-metal departments. We were very strong in hammer mills and stokers. Jerry Hahn and I got in on the ground floor of Gehl starting the forage harvester business. I

did a lot of fieldwork with Henry Gehl and his son, Ben Gehl, who was then chief engineer. We were making ensilage cutters [and] silo fillers, which were very popular in the foreign markets. They were making silo fillers and hammer mills during the war. Agriculture implement manufacture was a defense priority."[124]

WARTIME

As the war progressed, Gehl Brothers Manufacturing Company dealt with an increasing number of defense contracts and a decreasing number of workers. The West Bend-based company made numerous products for the war effort. As early as 1942, Gehl did work for Gibbs and Cox Systems for the United States Navy. The Company made bogie wheels for armored vehicles made by 4-Wheel Drive Company from nearby Clintonville, Wisconsin. Gehl Brothers Manufacturing Company also made welded gun carriages for anti-aircraft guns.[125]

Don and Lloyd Theisen, brothers from North Dakota who had started to work for Gehl Brothers Manufacturing Company during the Depression, recalled some of the war work the Company undertook in 1942 and 1943. "Gehl made bogie wheels in the foundry and machine shop," Lloyd Theisen said. "We made big drums for dryers to be used for drying clothes aboard ship. We made five molds a day of the ships' dryers. The shop made fifteen molds a day all told. The foundry machined the bogie wheels for tank treads on a grinding machine."[126]

By the early 1940s, Lloyd's brother, Don, was a foreman on the hammer mill line. The importance of the agricultural implement line was underscored by the difficulty both brothers had in enlisting for military service. "I had two six-month deferments because of my war work," Lloyd Theisen explained. Don Theisen only had one deferment, and he was the first of the two brothers to enlist in the U.S. Army on July 4, 1942. Lloyd enlisted in the U.S. Coast Guard on December 27, 1942.[127]

The Theisen brothers were typical of the hundreds of employees of Gehl Brothers Manufacturing Company, West Bend Aluminum Company, Enger-Kress, and other local companies that served their country during the war. Gehl and the local companies attempted to fill in as well as they could with women and teenagers from the West Bend schools, but labor shortages plagued the Wisconsin manufacturing economy throughout the war years.

For Gehl Brothers Manufacturing Company, the war was a welcome respite from the financial problems that had nearly

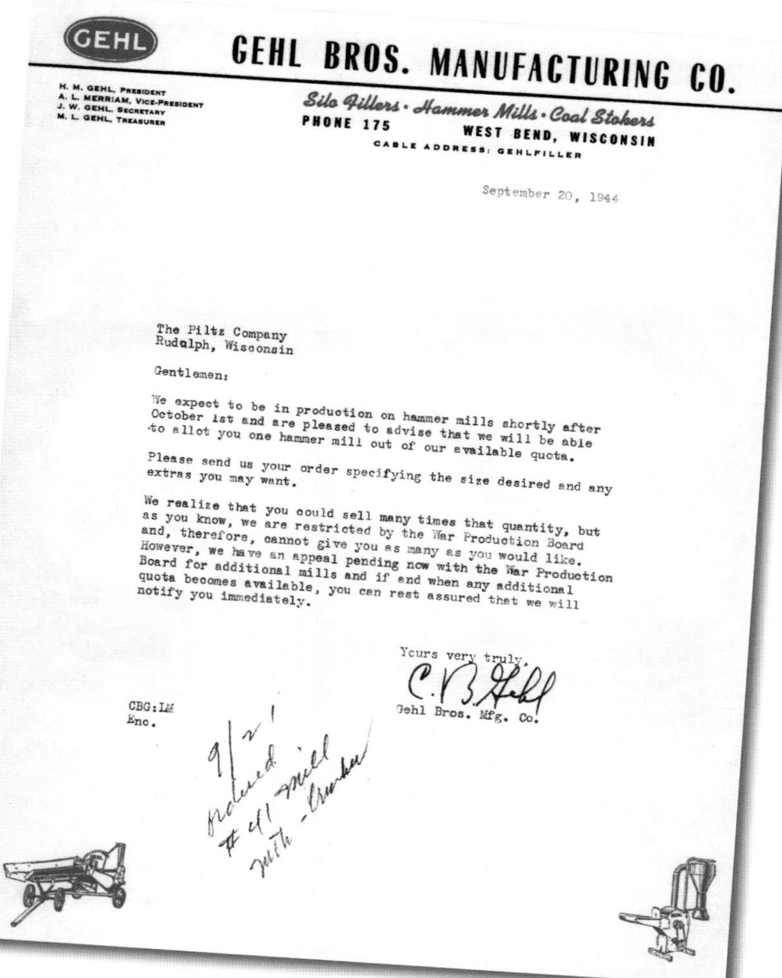

GEHL — GEHL BROS. MANUFACTURING CO.

H. M. GEHL, PRESIDENT
A. L. MERRIAM, VICE-PRESIDENT
J. W. GEHL, SECRETARY
M. L. GEHL, TREASURER

Silo Fillers · Hammer Mills · Coal Stokers
PHONE 175 WEST BEND, WISCONSIN
CABLE ADDRESS: GEHLFILLER

September 20, 1944

The Piltz Company
Rudalph, Wisconsin

Gentlemen:

We expect to be in production on hammer mills shortly after October 1st and are pleased to advise that we will be able to allot you one hammer mill out of our available quota.

Please send us your order specifying the size desired and any extras you may want.

We realize that you could sell many times that quantity, but as you know, we are restricted by the War Production Board and, therefore, cannot give you as many as you would like. However, we have an appeal pending now with the War Production Board for additional mills and if and when any additional quota becomes available, you can rest assured that we will notify you immediately.

Yours very truly,
C. B. Gehl
Gehl Bros. Mfg. Co.

CBG:LM
Enc.

An example of the manufacturing restrictions placed on Gehl Brothers Manufacturing Company by the War Production Board during World War II. Despite the impact of certain restrictions, during World War II the Company managed to remain on government priority lists for steel.

Following World War II, calendar art found its way into the nation's business advertising. Gehl was no exception, as shown by these illustrations that continued to appear in Gehl posters and advertisements.

felled the Company during the Depression. With strong agricultural implement business and guaranteed federal government defense contracts, Gehl Brothers Manufacturing Company was able to retire debt and begin planning for a profitable postwar era.

Second Generation

When the employees of Gehl Brothers Manufacturing Company began returning to the Company in late 1945, it was to a business that had essentially passed the baton of management from the first generation of Gehls to the second.

John, Mike, and Henry Gehl, who had managed the Company since early in the twentieth century, had increasingly delegated responsibility for the Company's management to the second generation, although the three older Gehl brothers still continued to come to work every day. By 1946, the four sons of John Gehl had parceled out management functions at the Company.

Although all four second-generation Gehl brothers were serving in officer capacities at the time, they were adamant in stating that there was no actual president or chief operating officer. The brothers were quick to downplay the importance of titles to maintain the feeling that everyone was equal.[128] None of the brothers used their titles; rather, they managed Gehl Brothers Manufacturing Company as a committee by mutual consent.

Albert Gehl recalled that early in the war, "Dad and his brothers were still active. Uncle Henry was the factory manager, and Mike was in sales. My dad was the manager."[129] During the war, several old-timers began disappearing from the scene, such as Arthur Merriam, who had handled the Company's promotional and advertising accounts since 1911. John Schwinn was another old-time employee, whose tenure in the Sales Department dated to the 1920s. Henry Arnfield remained at the Company as the office manager and assistant treasurer, a role he continued as B. C. Ziegler's proxy for the reorganized company.

One of Henry Gehl's sons, Henry, Jr., served as the plant superintendent in the years just after the war. That all changed at the end of the 1940s, when John's sons took over. They brought with them a management style that relied on the common touch.

Don Theisen spent nearly a year and a half in occupied Europe with the U.S. Army before returning to West Bend in late 1945. Don said that he knew the second generation of Gehls better than he did the first. In the immediate postwar years, Don worked closely with Mark, and he said that Mark

The first and second generations of Gehl brothers examining a Gehl model FH48 forage harvester. From left to right are: Dick, John, Mark, Mike, Al, Henry, and Carl. *(Photo courtesy of the* Milwaukee Journal Sentinel*)*

made the Gehl Company. "Albert handled the finances and he never threw anything away," Don Theisen said. "Mark always carried half-dollars in his pocket, and you could hear them jingling. Mark was Beep No. 1 on auto-call [paging system]."[130]

The second generation of Gehl brothers. From left to right are: Carl, Dick, Mark, and Al.

In 1947, Don Theisen was president of the old AFL UAW local at Gehl Brothers Manufacturing Company. He recalled that the second generation of Gehls never lived like they had money. One time, Don Theisen drove Mark Gehl's Plymouth to Sheboygan Falls and remembered it as "a dilapidated piece of machinery. But," continued Don, "if you met Mark Gehl at a tavern, you never had to pull out your billfold, because he never let you buy a drink."[131]

Dick Gehl was "on the road" much of the time, and the stories about his generosity were legion. "Dick went up north with a salesman," Don Theisen said, "and they went to Mass on Sunday. Dick had a $5 bill out for the collection and he also put a $20 bill in the building fund collection."[132]

Lloyd Theisen recalled an incident involving Dick Gehl. Dick called Lloyd into his office and told him that he had talked to Lloyd's father in North Dakota. Dick explained that he had been traveling in western North Dakota and stopped to see an implement dealer in Beach, Lloyd's and Don's hometown. He told the dealer that Gehl employed a couple of young men from Beach, and the dealer asked Dick if they were the

{ ALBERT }

Albert Gehl has been alive for two-thirds of Gehl Company's one hundred and fifty years. Albert, who served in various officer capacities for the Gehl Brothers Manufacturing Company from 1945 to 1972, including as its president, celebrated his ninety-ninth birthday in August 2008.

Albert Gehl was born in West Bend on August 17, 1909. Teddy Roosevelt was the nation's outgoing president, and Robert LaFollette was an up-and-coming politician in the Badger State. The U.S. Government minted the first Lincoln-head pennies, and in far-off Panama, workers started pouring the first concrete for the canal that would span the Panamanian Isthmus. Robert Peary planted the U.S. flag at the North Pole, and ocean liners were able to cross the Atlantic Ocean in just over four days.[141]

Michael Gehl, Albert's grandfather, died just after Albert was born, but Albert recalls visiting the family farm at St. Lawrence as a boy. The Gehl family spoke some German in the home, but Albert's father, John, encouraged his children to speak English as much as possible. Still, Albert spoke German in school at Holy Angels Roman Catholic Elementary in West Bend, where he was taught well by the School Sisters of Notre Dame.

Being Catholic, the Gehl family observed meatless Fridays, and Albert has vivid memories of eating navy bean soup each Friday. Being good Luxembourgers, "we made our own sauerkraut in a big crock. Dad made wine. West Bend had its own brewery. When beer came back in 1933, we stayed up until past midnight so we could go to the brewery and get a case. I gave my brother Frank a bottle of near beer as a joke. He said it tasted just like the stuff we drank during Prohibition. We howled."[142]

Born in 1909, Albert remembered that the family always had a car. "We had a Ford Model T," he said. "Buick made a seven-passenger car, and the dealer had it on the lot for a long time. Dad bought it, and he told Frank to drive into the garage. Frank ran the car into the garage and tore the door off the Buick."[143]

Although brothers Mark and Dick went to Campion Preparatory School—the Jesuit preparatory school in Prairie du Chien, Wisconsin—Albert, along with his brothers Frank and Carl, graduated from West Bend High School. "They didn't have a gymnasium until 1927, the year I graduated," Albert said, "and then a new high school on Main Street was built. The old high school was kitty-corner from the Holy Angels Rectory. The building is still there."[144]

Albert Gehl, left, presenting the Richard M. Gehl Outstanding Performer Award, named in memory of Albert's late brother, to salesman Gerald Luckert.

After his graduation from Marquette, Albert served his country in the U.S. Coast Guard during World War II. He was stationed on the East Coast in 1942 and then spent twenty-six months as a Navy Reserve officer monitoring the big Pratt & Whitney plant in Kansas City, Missouri. "I would see Harry Truman when he was vice president," Albert said. "I was on the elevator with him more than once."[145]

Albert met his wife, Lorraine, in the service, and they were married on Memorial Day 1945 at St. Sebastian Parish in Milwaukee. Daughter Kathy was born in 1946 and daughter Carol was born two years later.

Albert spent his entire career with Gehl Brothers Manufacturing Company and held the positions of general manager, secretary, and vice president before eventually becoming president in 1969. Albert said he served as president only because the other officers at the time were older and closer to retirement.[146]

Albert Gehl remained president of the Company until 1972, when Joe Zadra succeeded him in the office. He served another four years as chairman of the Board and retired in 1976. Today, he lives at the Milwaukee Catholic Home, a living reminder of the Gehl institutional and family memory.

After passing on his sales manager duties to his nephew, Dick Gehl (left), Mike Gehl (right) remained active on the Gehl Board of Directors and in civic activities in West Bend after retiring from management. Mike Gehl served as mayor of the city from 1948 to 1954. *(Photo courtesy of* Northwest Farm Equipment Journal*)*

Richard ("Dick") M. Gehl, 1901–1963

Theisens. When the dealer told Dick that their father worked at the lumberyard at the edge of town, Dick drove down to say hello and tell him of the latest news of his sons.[133]

Expanding the Base

Gehl Brothers Manufacturing Company began to expand its business in the immediate postwar years. Ten years of the Depression and five years of war had created a pent-up demand for capital goods, both in the consumer economy and on the farm. Consumers wanted new cars, and farmers wanted new agricultural implements.

"World War II built up a big backlog of demand for farm machinery," a New York investment banking firm reported about the agricultural implement industry in 1947. "The industry, with its large war products output, was unable to make enough farm equipment to meet the tremendous war requirements for food. With the available farm machinery in use at maximum capacity, much of the existing equipment today is either over-age or worn out."[134]

In 1937, farm equipment sales nationwide were $500 million. The report estimated that beginning in 1948, farm equipment sales would easily double that. The wartime boom left American farmers with a capital surplus available for investment in new equipment. By 1946, cash farm income in the United States was $20 billion, more than double the figure reported in 1937.

By 1947, Gehl was gaining a reputation for making top-quality forage harvesters, which had been introduced during the war. In a 1948 *Milwaukee Journal* article, John Gehl explained that the Company's forage harvester replaced the ensilage cutter, corn binder, and hay loader. The Company's dollar volume of business in 1948 was five times what it had been in 1940. Business was so good, John said, that the Company had discontinued its coal stoker line, which had first been introduced in 1931.

The Gehl Brothers Manufacturing Company plant had closed its foundry in early 1948 but had added several warehouses in the immediate postwar era to handle increased inventory. The plant employed about 400 workers year-round, double the employment figures from before the war. Gehl Brothers Manufacturing Company ramped up production seasonally and hired local farmers during the winter to build inventory for the next spring.[135]

John Gehl noted that the Company adhered to a business philosophy that "the small manufacturer either has to furnish the best machine or sell it for less money."[136] Gehl Brothers Manufacturing Company had decided to pursue the "best machine" approach, John Gehl told the *Journal* reporter.

In the immediate postwar era, the bulk of Gehl Brothers Manufacturing Company's sales were in Wisconsin, Minnesota, and Iowa. With the growth in demand, however, the Company quickly increased its sales force in the period between 1948 and 1952. Gehl Brothers Manufacturing Company soon had a presence in all forty-eight states and most Canadian provinces. In 1949, the West Bend-based implement manufacturer began exporting machinery and began licensing implement technology overseas; Gehl Brothers Manufacturing Company soon had customers in thirty-five foreign countries.

Gehl Brothers Manufacturing Company enjoyed amicable relations with its union work force, and in 1948, the Company became a closed shop. Employees had to pay three dollars to join the union, and union dues in the late 1940s and early 1950s were one dollar a month. Although Carl Gehl was in charge of labor relations, most grievances were settled over a glass of beer and a game of cards with Mark Gehl, the plant manager.

The above aerial shows the Gehl plant in 1948, the same year the foundry closed. The below aerial shows the plant's transformation by 1957.

Top: The first forage harvester attachments were the hay pick-up and the one-row crop attachments.

Right: For many years, the Company emphasized the simplicity of operating its equipment. It most commonly declared: "Any boy can do a man's job," as shown on this forage harvester advertisement. Other Gehl catchphrases included: "Two horses can pull it," and "The hired man won't break it."

Testing

Gehl Brothers Manufacturing Company's increased presence in the United States farm marketplace meant that the Company had to send service technicians from West Bend to dealers across the country to set up and repair equipment. Harold Gauger was twenty-three years old when he started working on the assembly line on November 24, 1947, at ninety cents per hour making hammer mills. "I put cylinders together on the side of the line," Gauger recalled. "I worked in the old building, which was old horse barns. I also did coating for overseas shipping. We had one forklift with three wheels. It had a single air-cooled engine and no seat. I stood on the back."[137]

Gauger first became acquainted with the Company's Test Department in 1949. He recalls that "it was ten degrees below zero when I went to California to rebuild mower bars that year. I took the train to Chicago's Union Station and then a limo to Midway Airport. O'Hare Airport wasn't there yet. I flew TWA on a DC-7, four-engine plane from

Gehl was the first to manufacture a PTO two-row chopper. The Chop-All harvester appealed to farmers because its low cost meant spending the same amount on a two-row chopper as they would on a one-row.

A long-standing tradition of Gehl Company has been to distribute a gift of turkeys or hams to its workers during the holiday season. This custom dates back as early as 1937, as shown here, and continues today with the annual distribution of turkeys at Christmastime.

Top: The Gehl self-unloading forage box made its first appearance in 1955 and remained a significant product in the Gehl line for over fifty years.

Left: A Gehl Brothers Manufacturing Company sales truck of the era.

On-the-farm feed processing allowed farmers of all types to cut costs, save time hauling feed, and have flexibility by being able to mix on their own schedules.

Chicago to Bakersfield, California. I was in the air for eight hours. I was wearing a cap and boots and everybody looked at me. The dealer was in Fresno. I was out there six weeks rebuilding mower bars. When I came back, my one-year-old son said, 'Who's that?'"[138]

Gauger came back to West Bend, was promoted to foreman, and spent four years, 1952 to 1956, as a sub-assembly foreman. In 1956, he joined the Test Department as an engineering technician, which took him to forty-six states as well as Canada, Japan, Ireland, England, and Mexico.

While he was in Department M, Gauger was witness to growing technological change at Gehl Brothers Manufacturing Company. In 1954, Gehl added multiple-row crop attachments to its forage harvester line and introduced a self-unloading forage box the next year. In 1959 and 1960, Gehl Brothers Manufacturing Company developed a silo-based re-cutter, the all-purpose Gehl Mix-All grinder mixer, and a seventy-two-inch flail chopper.

The development of power take-off (PTO) equipment in the 1930s revolutionized agricultural implements. A PTO was a splined driveshaft, typically on a tractor, that could be used to provide power to an attachment or separate machine. Designed to be easily connected and disconnected, the PTO allowed implements like the forage harvester to be powered from a tractor or other power source.

Although Gehl Brothers Manufacturing Company developed a PTO-driven manure spreader in the late 1930s, it was never put into production. The Company's entry into the PTO-driven machinery field began with forage harvesters in 1943. Gehl Brothers Manufacturing Company purchased its PTO components from local manufacturers, including the Pick business in West Bend and Weasler Engineering and Manufacturing located north of West Bend.

"The Pick family had two plants in West Bend," explained Gene Wendelborn. "They took over the Schmidt & Stork Wagon Company plant that made replacement auto parts. Originally, Pick made the PTO equipment for Gehl. They started to make farm-unloading boxes. Tony Weasler was Pick's chief engineer. Weasler started his own PTO plant on the outskirts of West Bend. They made PTO equipment for Gehl Brothers Manufacturing Company. In fact, they were very much connected with the Company after 1950."[139]

Gehl made the most of tractor power through direct PTO drive. The Gehl Hi-Throw PTO blowers were able to easily fill silos up to eighty feet tall, and do it fast, at over sixty tons of grass or corn silage an hour.

The Office Staff

Mark Gehl had a hard-and-fast rule for managing office overhead costs. He told colleagues in the 1950s there should be no more than one office staffer for every ten employees on the manufacturing line. For much of the 1950s, when Gehl Brothers Manufacturing Company employed about 400 full-time workers, there were thirty-five to forty people in the Company's West Bend offices.

The second generation of Gehl brothers at work in the office. One distinct feature of the offices during this time was a door connecting one office to the next, allowing one to walk from office to office without stepping into the hallway.

Laura Heisler joined the Company in 1948 as a keypunch operator in the Payroll Department. Heisler had grown up in St. Killian and West Bend and was a recent West Bend High School graduate when Henry Arnfield hired her. Her father had worked in the foundry for years, and her brother worked in the Company's Inspection Department.

When Heisler started at Gehl Brothers Manufacturing Company, the office staff used manual upright typewriters and inserted onionskin and carbon paper when they wanted to make copies. Heisler recalled one incident in the summer of 1948 when the foundry, which was in the process of closing, was belching smoke. The office staff had the windows on the second floor wide open to capture any breeze available in the stifling mid-afternoon heat. Suddenly, a spark from the foundry ignited a pile of papers on the office manager's desk. The office staff quickly stomped out the resulting blaze.

The accounting office used heavy, crank-style adding machines and two- and three-post Boston Ledgers. It was not until the late 1950s and early 1960s that the office staff was outfitted with new, IBM Selectric typewriters.[140]

The 1940s and 1950s were a time of growth and prosperity for Gehl Brothers Manufacturing Company. The Company increased its product line and offerings, strengthened its financial base, and invested in research and development. The 1960s and early 1970s would lead to even more growth.

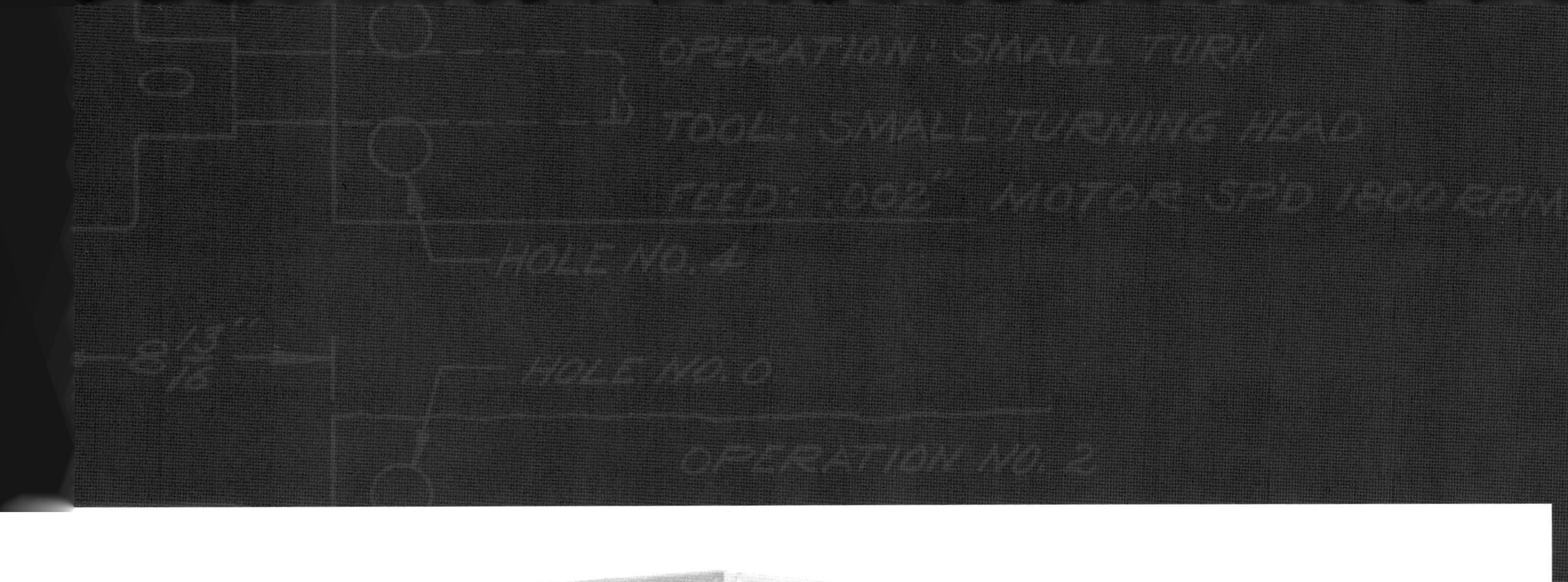
OPERATION: SMALL TURN
TOOL: SMALL TURNING HEAD
FEED: .002" MOTOR SP'D 1800 RPM
HOLE NO. 2
HOLE NO. 0
OPERATION NO. 2

GEHL 2600

Chapter Five

GEHL *Company* 1961–1976

In the 1960s, much was new for Gehl Company: new name, new products, new management, and even new colors.

The new name came in the summer of 1967 when the Company announced that it was changing its official name from Gehl Brothers Manufacturing Company to Gehl Company. In a statement to shareholders, management noted that "the new name more typically describes this firm both now and in the future. We believe that the promotion of the Company and its products is enhanced by a short, impact-carrying company name."[147] Regardless of the Company's official name at the time, one can still hear longtime West Benders and area residents refer to the even shorter version, "Gehl's."

The new name was accompanied by a new color scheme for Gehl products. Since Charles A. Silberzahn began producing Hexelbank ensilage cutters in the 1890s, the Company colors had been a rust red and drab olive green. In late 1966, Gehl announced a new color scheme for its products. The new colors were officially called "blaze red" and "maize yellow." Joe Ecker, the Company's sales manager, said, "The wide availability of TV and other public media have made today's farmer as aware of design showmanship as his big-city counterpart."[148]

Changes At the Top

The 1967 name change was recognition that the first two generations of Gehl brothers were fading from the scene. In 1967, Michael Gehl and Henry Gehl were Board chair and president of the Company, respectively, although both were long retired from active management. John W. Gehl's sons, Mark, Albert, and Carl, managed the day-to-day affairs of the Company. Albert was serving as general manager and secretary. When his uncle Mike died in 1969, Henry became Gehl Company's chairman of the Board. The family then asked Albert to take the position of president. Mark Gehl served as vice president of manufacturing throughout the 1960s and early 1970s, and Carl Gehl handled labor, legal, and human resources issues as the Company's vice president and export manager.

Besides the loss of Mike Gehl in 1969, Gehl suffered two other management losses early in the 1960s. After nearly three decades as the Company's financial manager, Henry Arnfield retired from Gehl Brothers Manufacturing Company in 1963. Late that year, the Gehl community and West Bend were stunned when Dick Gehl was felled by a massive heart attack while watching a Green Bay Packers game on television at his home on Big Cedar Lake.

Dick Gehl had been the public face of Gehl Brothers Manufacturing Company since the 1920s. As sales manager for the Company, he knew literally every implement dealer who handled Gehl equipment, from one end of the United States to the other. Dick Gehl was only sixty-two when he died.

"Dick was the sales manager and he was a tremendous sales manager," recalled Joe Ecker, his replacement. "He and his wife, Toots, traveled the world, and he knew every dealer in the United States. He was a very positive person, an outgoing and fun-loving person. Dick was the 'outside' man. Every winter, Dick would drive his car to California. He would stop to see dealers across the country. His trademark was a Stetson hat that he wore. He was a big man, and he was very gregarious."[149]

Dick Gehl was the first of the second-generation Gehl brothers to die, and his death forced the remaining brothers to start looking at a plan of succession. Their choice to replace their oldest brother was Joe Ecker, who came to Gehl Company in 1958 after graduating from the University of Wisconsin–Madison. Ecker first went to work as an advertising manager for the Brillion Iron Works, a foundry and farm machinery company in his hometown. "I came to Gehl via an approach from the Gehl brothers," Ecker said. "I was at the Brillion Iron Works from 1950 to 1957 in their agricultural works. Brillion was not a competitor since they made tillage tools only."[150]

Another key executive promotion involved Joe Zadra, a ten-year veteran, who was named to replace Henry Arnfield in 1963. Zadra was born in an iron-mining region near Ironwood, Michigan, and was raised on a dairy farm in Wausau. He attended the University of Wisconsin. After graduation in 1948, he joined an accounting firm in Green Bay and traveled extensively in the Upper Peninsula. "I was introduced to data processing with punch cards," Zadra said. "I joined Gehl in 1952, and I had been in accounting from 1948 to 1952. When I came to Gehl, I handled a tax case with the Wisconsin Department of Revenue. It was an accounts receivable bad-debt-reserve issue. I came in as office manager and controller under Henry Arnfield and later became treasurer and vice president of finance."[151]

Giving Farmers a Hand

In 1968, Mark, Albert, and Carl Gehl sat down for an extensive interview with a reporter from the business desk of the *Milwaukee Sentinel.* Gehl Company was in the midst of a major expansion at the time, and the brothers noted that the current Gehl factory in West Bend occupied 325,000 square feet along a three-square-block area fronting Water Street. The Gehl complex included an additional 16,000-square-foot

This 1972 aerial photo shows the addition of the east plant at the very top of the picture.

building for the Engineering Department and a 70,000-square-foot warehouse completed in 1966.[152]

The Gehl brothers told the reporter that the first phase of a 440,000-square-foot expansion of the Company—essentially doubling the production and storage capacity of the firm—was underway on a thirty-three-acre tract just east of the Gehl plant. The first 140,000 square feet of expansion space, which was destined to be used for fabricating and machining, was slated to be completed before the year's end in 1968. The remaining 300,000 square feet of expansion facilities would be brought on-line during the next ten years.[153]

In 1968, Gehl Company was becoming one of the larger manufacturing firms in the Milwaukee metropolitan area. Although nowhere near the size of such giants as Allis-Chalmers, Harley Davidson, and Briggs & Stratton, Gehl had grown from approximately one hundred employees twenty years before to a manufacturing powerhouse that employed 1,000 people working three shifts. The Company's annual payroll of $38 million had a major economic impact in West Bend and surrounding Washington County.[154]

Gehl Company's growth was predicated on the increasing mechanization of America's farms. In the years after World War II, the number of farms began a decline that has not abated to the present day. There were fewer and bigger farms, and those farms made more money than farms had in the past. Mechanization meant that the farmer could operate a farm with fewer hired hands. Real farm gross domestic product (GDP) per person in agriculture in the United States had averaged 1-percent growth per year from 1880 to 1940. By 1960, the annual growth of farm GDP was almost 2.8 percent.[155]

New and improved tractors, pneumatic tires, hydraulics, silage and hay choppers, forage harvesters, multi-row planters, portable feed mixers and grinders, and dozens of other technological advances transformed farming in the years after World War II. As new technology appeared in dealer showrooms in the 1960s, it hastened the trend toward larger farms and fewer farmers, and it made the American farmer among the most productive on earth.

Nothing signified the dominance of American mechanized farming as much as Soviet Premier Nikita Krushchev's visit to the Coon Rapids, Iowa farm of hybrid seed promoter Roswell Garst in the fall of 1959. The Russian premier—himself a native of a farm in the Ukraine—and Garst spent several hours

The first Hexelbank Club meeting was held July 31, 1967, at The Linden Inn, to honor those employees with twenty-five or more years of service at Gehl. Sixty members were inducted at the first meeting.

A Gehl self-unloading forage box unloading into a Gehl Hi-Throw blower, filling a silo on the Richard Stoffel Farm in 1966. The Company placed emphasis on the fact that this machinery required little effort on the part of the farmer and lessened the need for hired help.

discussing the merits of tractors and tillers and seed corn.[156] Although America and the Soviet Union were engaged in a Cold War that would not end for another twenty years, the two superpowers understood that their power would wane quickly if they could not feed their people.

New Products

Gehl followed its introduction of innovative forage harvesters during World War II with the establishment of a dealer network to sell the new machines to farmers across America. Gehl was not the only American manufacturer of forage harvesters, but its early entry into the field gave Gehl name recognition among farmers. And forage harvester sales literally skyrocketed between 1945 and 1964. U.S. farmers purchased 20,000 forage harvesters between 1942 and 1945. Ten years later, there were ten times that number in operation on American farms, and by 1964, in less than twenty years, U.S. manufacturers had sold some 316,000 forage harvesters.

Much of Gehl Company's success was due to targeting a niche market in the sometimes cutthroat business of making farm implements. In the 1960s, Gehl specialized in forage harvesting and forage handling equipment. The Company's net sales volume was $10 million at the beginning of the decade. The markets were very dairy farmer-oriented and stayed that way. The size of the dairy farm in those days was quite small compared to the size of farms today. A large farm in Wisconsin was 120 acres and consisted of twenty milking cows. There were a lot of farms that were as small as eighty acres. "These were all family farms," explained Joe Ecker. "One farmer was buying a forage harvester, a forage blower, and a self-unloading forage box."[157]

Gehl had a field sales force, and it sold product through farm equipment dealers in every small town. "We sold exclusively through implement dealers," Ecker recalled. "We had six salesmen in the state of Wisconsin who called on a couple of hundred dealers. Regional sales managers in West Bend supervised them. We grew that sales force as we went along."[158]

Gehl also targeted farm journals such as *Successful Farming, Farm Journal,* and *Capper's Farmer,* and state publications such as *Wisconsin Agriculturist, Indiana Farmer, Prairie Farmer,* and *Pennsylvania Farmer.* "We would send them ad slicks and insertion orders," Ecker said. "The Brady Company was our ad agency. They were located in Appleton, Wisconsin, and then moved to Milwaukee. The account executives would come here, and we would create the ads. Our advertising budget was probably a couple hundred thousand dollars in the late 1950s, early 1960s. We had a secretary and an assistant ad manager, and that was it. And the secretary worked for a lot of other people, too."[159]

Gehl would use its semi-annual dealer open houses to introduce new products, and dealers had a lot to celebrate in the 1960s. In 1960, the Company's dealers had been the first to

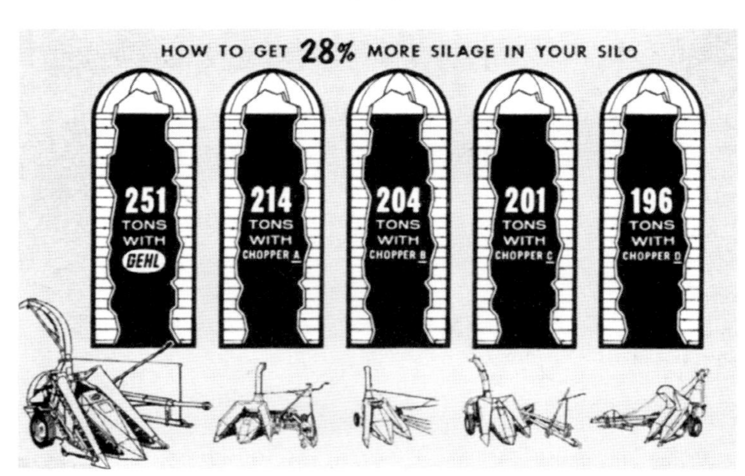

This chart was used by Gehl Company to show the amount of silage a sixteen-by-fifty-foot silo would hold when filled with silage cut by a Gehl chopper as compared to four other leading choppers. Tests were conducted by the Brady Company at the 1962 Farm Progress Show.

A small-town company in the big city. Gehl held a national dealer meeting in Las Vegas in 1976. The Company's marketing scheme at the time was the "Showdown." The idea was to encourage customers to compare Gehl machines on features, performance, and value to other brands. Gehl "challenged" all other brands because it believed its machinery was the better buy.

Gehl products have been marketed around the world for decades. Throughout the years, Gehl has relied on its own marketing team, its dealers' selling effort, and at times, ad agencies to help promote the Company's name. Shown here, attending a 1984 marketing meeting, are (from left to right): Karl Olhm, head of Brady Company; Carroll Merry, Gehl advertising director; John Gehl, Gehl vice president; Joe Ecker, Gehl director of marketing; and John Smithers, Gehl director of sales.

preview the new Gehl flail chopper, the first on the market to boast a six-foot chopping width. At a dealer open house in 1964, Gehl introduced its new Chop-King forage harvester, which it called "the biggest chopper going" in pull-type forage harvesters.[160]

The Company also upgraded its Chop-All forage harvester line with the introduction of a variable-cut transmission and an easy-swing drawbar on the popular pull-type models.[161]

That same year, Gehl introduced its line of rotary shredders and a pick-chop attachment for forage harvesters that allowed the farmer to pick only the ears from one row of corn while gathering in the entire stalk from the other row. In early 1965, the Company debuted what would become its best-selling Mix-All feed maker with a fiberglass mixing tank, the first company in the industry to do so. Gehl initially had to sell the concept of on-farm feed making, but the concept took hold and proved very successful for the Company. It advertised that fiberglass would not rust, dent, or corrode.[162] The Gehl Mix-All grinder-mixer became one of the Company's most significant product lines. At its peak in the late 1960s and early 1970s, the West Bend plant was producing approximately 240 Mix-All units every week.[163]

In the summer of 1967, Gehl developed a lucrative new market when it began to introduce the first of a series of accessories for its forage harvesters. "Gehl pioneered the narrow-row corn head attachment for forage harvesters," *Drovers Journal* explained in its 1967 annual roundup of farm machinery.[164]

That fall, the Company introduced a nine-foot mower that "conditioned" (or crushed the crop stems so they dried faster) and windrowed a full nine-foot swath. *Prairie Farmer* noted that the new machine "operates with four heavy-duty roller chains for low maintenance. Floating-head design with

Above: The Gehl "72" flail chopper was the year-round answer to a wide variety of cutting and shredding jobs. Most of all, the chopper more accurately controlled vegetative cover.

Right: Gehl strived to develop machines that produced and harvested high-quality forage. Providing an even, short cut helps preserve more nutrients by eliminating room for oxygen from the silo. The Chop-All harvester featured a Select-A-Cut transmission, which allowed the operator to choose from a one-fourth-, three-eighths-, or three-fourths-inch forage length.

skid shoes at both ends allows the machine to follow ground contour without digging in."[165]

The late-fall introduction of the model RC800 cylinder re-cutter made 1967 one of the most prolific in the Company's long history. With nine knives, the twenty-two-inch-wide, twenty-four-inch-diameter RC800 brought nearly 1,000 square inches of screen area to the silo for the re-cutting of high- and low-moisture corn, haylage, silage, bale slices, cornstalks, and other roughage. Gehl advertised that the unit easily attached to the model FB88 Hi-Throw forage blower and that the model RC800 included a self-contained knife sharpener, apron feed chain, and double feed rollers.[166]

In 1969, Gehl introduced a number of refinements to its product lines, including a new six-knife model CB600 cylinder-screen forage harvester equipped with separate blower for loading the box;[167] a new model CT300 cut-and-throw cylinder forage harvester with nine knives;[168] and a blender feed box suitable for fenceline and bunk feeding, silage, green feed, hay, corn, ground feed, or roughage hauling.[169]

Farm journals reported extensively on the latest Gehl innovations, including the 1969 introduction of a new self-unloading forage box featuring a two-speed worm-gear drive shaft.

Having a Mix-All on the farm allows farmers to control the ingredients put into the feed, especially when adding vitamins, minerals, or drug pre-mixes to farm-produced mixed feeds.

Feed making flexibility was made possible by the Gehl Mix-All with the proper attachments and a full set of screens. Small pig farmers were able to produce the fine grinds necessary for small pig rations, while coarser grinds for swine and dairy and beef cattle were achieved simply by changing the screen size.

By the 1960s, Gehl products had entered the international market. The photo on the left shows the model 300 and 600 forage harvesters being introduced in France in the late 1960s. The bottom photo, taken in the fall of 1969, shows salesmen and mechanics of the Gehl distributor for Spain putting a mower conditioner to use in a field near the town of Manzanares, Spain.

Gehl innovations went beyond farm equipment. In 1970, the Company introduced the Gehl Blazer, an all-season vehicle with skis for winter and wheels for summer, and with dune-buggy applications. During a time when many equipment manufacturers were developing snowmobiles, the development of the Blazer did not come as much of a surprise. It was, however, discontinued a short while later.

In 1968, Gehl made the switch from flywheel- to cylinder-type harvesters. Cylinder-type re-cutters and silo fillers were manufactured by Gehl as early as 1917. The Gehl model CB600 cut-and-blow harvester introduced a special air feed-flow opening below the cutting mechanism that quickly transferred material from the cutting cylinder housing into the blower.

The Gehl model RC800 re-cutter provided year-round versatility and the convenience of performing fine chopping right at the silo.

The Gehl model BU810 self-unloading forage box made for smooth, steady, and safe unloading. A safety bar was located both across the box front and above the discharge.

National Livestock Reporter commented on the safety features built into the new Gehl equipment: "Set-back top beater runs slower than lower beaters so that there is no slug delivery to the front conveyor. All running parts are shielded; it also has safety bar protection. The cross conveyor bottom is enclosed."[170]

New and improved product lines continued to come out of Gehl Company's West Bend Research and Development Department in the 1970s. The department was headed by Don Burrough, who was recruited from a major competitor and served as vice president and director of engineering. In 1970 and 1971, the Company brought out a 285-cubic-foot-capacity three-auger mixer-feeder box with optional hydraulic drive and an electronic weigh scale, as well as its model MS410 manure spreader, which held 335 bushels, or seven and a half tons.[171] In 1975 and 1976, the Company introduced the model CB800, its largest pull-type forage harvester, and its first totally hydraulic grinder-mixer, respectively.

Gehl Company's innovations in the 1960s and 1970s positioned the West Bend implement manufacturer to be a leader in forage harvester development and manufacture. But it was a crowded field. In the 1970s, numerous manufacturers in the U.S. and Europe were competing in the forage harvester marketplace, including John Deere, New Holland, Case, Massey Ferguson, Oliver, Claas, and Hesston.

In 1970, Gehl made the decision to diversify its product lines, a step that would ultimately change the Company. The Company began licensing and later manufacturing skid-steer loaders.

First Foray into Construction Equipment

Skid-steer loaders were new versatile four-wheel-drive, rubber-tired machines that could turn within their length. Equipped with a bucket or fork tines, they could be used in tight spaces to feed livestock, clean out barns, and lift and load heavy payloads.

One of the world's first skid-steer loaders was developed by a northwestern Minnesota turkey farmer as a three-wheeled machine in 1957. The next year, the Melroe Manufacturing Company of Gwinner, North Dakota, purchased the manufacturing rights to the machine. Melroe, which had been formed ten years earlier to manufacture windrow pickups for hay farming, brought out a four-wheel model of the skid-steer loader and branded their new product the Bobcat in 1962.[172]

During the next ten years, the Bobcat and similar skid-steer loader models found increasing application at farms and construction sites in North America and Europe. Albert Gehl, then-president, told the executive team that Gehl Company needed to diversify.[173] The result was the first Gehl skid-steer loader.

Gehl Company did not initially manufacture the new line. It was manufactured by the Ericsson Company in Minneapolis, Minnesota, and private-labeled for the Company. Ericsson was also manufacturing it for Ericsson's own use and for Ford Motor Company. In 1972, Gehl started buying smaller skid-steer loaders, the model HL2500, from Hydra-Mac, Inc., of Thief River Falls,

The Gehl self-unloading forage box was easily adaptable for truck mounting, which was useful in bunk feeding or unloading into the blower or elevator.

Gehl Company's Madison, South Dakota employees in 1973, the year Gehl opened the new plant to produce skid-steer loaders and round balers. Fran Janous, who headed the Madison Plant for many years, is kneeling in the left foreground.

Gehl Company's promotions included a free gift, such as a bicycle, Christmas tree, or cookware, for customers making an equipment purchase.

The 2000 series Gehl skid-steer loaders were economically priced, yet included many options normally available on only the largest skid-steer loaders.

{ THE STRIKE }

For Joe Zadra, the nine-week work stoppage by the Allied Industrial Workers against Gehl Company in the spring of 1973 was "a real challenge to my years as president."[179] Zadra had been elevated to the presidency in the spring of 1972, when Albert Gehl stepped down from the position to succeed his uncle Henry as chairman of the Board.

With the exception of minor work stoppages, the Allied Industrial Workers had been representing Gehl workers since an organization fight in 1938 and a several-week strike in 1941. Originally chartered as the United Auto Workers (UAW), the locals in the Milwaukee area remained with the American Federation of Labor (AFL) when the AFL and the Congress of Industrial Organizations (CIO) split in 1937. When the AFL and CIO reconciled in 1955, the Detroit locals of the UAW were admitted back into the AFL-CIO. The next year, the Wisconsin locals of the UAW reconstituted themselves as the AIW.[180]

For Gehl Company, the 1973 strike was a surprise. Mark Gehl, who served as the Company's vice president of manufacturing and engineering from 1957 until his 1970 retirement, had negotiated factory-floor labor issues with the AIW local since the end of World War II. He was well known for his casual negotiating style through the 1950s and mid-1960s, at which time Carl Gehl took over labor issues. But Mark Gehl was gone by 1973, and casual negotiations were as well.[181]

Negotiations for a new contract opened early in 1973 and rapidly reached an impasse. On March 31, more than 700 members of AIW Local 579 voted by a wide margin to reject Gehl Company's final offer.[182] Picket lines went up on April Fool's Day, and both sides prepared for a long strike.

In the first three weeks, Gehl Company and the union met only twice, and that was at the urging of the Federal Mediation and Conciliation Service.[183] The two sides were remarkably tight-lipped about the issues involved, although wages were at the heart of the dispute. The Company's final offer in late March had been a raise of fifty-five cents an hour over two years; the union demanded a full dollar.[184]
The strike dragged on through May, and not until early June did the two sides narrow their differences.

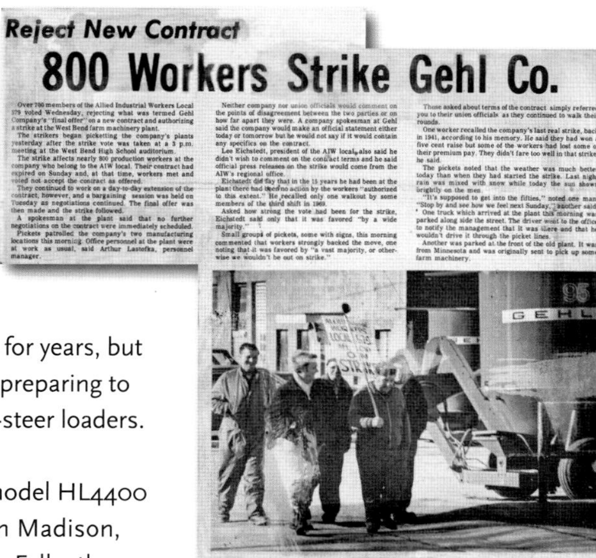

Reject New Contract

800 Workers Strike Gehl Co.

Over 700 members of the Allied Industrial Workers Local 579 voted Wednesday, rejecting what was termed Gehl Company's "final offer" on a new contract and authorizing a strike at the West Bend farm machinery plant.

The strikers began picketting the company's plants yesterday after the strike vote was taken at a 3 p.m. meeting at the West Bend High School auditorium.

The strike affects nearly 800 production workers at the company who belong to the AIW local. Their contract had expired on Sunday and, at that time, workers met and voted not accept the contract as offered.

They continued to work on a day-to-day extension of the contract, however, and a bargaining session was held on Tuesday as negotiations continued. The final offer was then made and the strike followed.

A spokesman at the plant said that no further negotiations on the contract were immediately scheduled.

Pickets patrolled the company's two manufacturing locations this morning. Office personnel at the plant were at work as usual, said Arthur Lastofka, personnel manager.

Neither company nor union officials would comment on the points of disagreement between the two parties or on how far apart they were. A company spokesman at Gehl said the company would make an official statement either today or tomorrow but he would not say if it would contain any specifics on the contract.

Lee Eichstedt, president of the AIW local, also said he didn't wish to comment on the contract terms and he said official press releases on the strike would come from the AIW's regional office.

Eichstedt did say that in the 15 years he had been at the plant there had been no action by the workers "authorized to this extent." He recalled only one walkout by some members of the third shift in 1969.

Asked how strong the vote had been for the strike, Eichstedt said only that it was favored "by a wide majority."

Small groups of pickets, some with signs, this morning commented that workers strongly backed the move, one noting that it was favored by "a vast majority, or otherwise we wouldn't be out on strike."

Those asked about terms of the contract simply referred you to their union officials as they continued to walk their rounds.

One worker recalled the company's last real strike, back in 1941, according to his memory. He said they had won a five cent raise but some of the workers had lost some of their premium pay. They didn't fare too well in that strike, he said.

The pickets noted that the weather was much better today than when they had started the strike. Last night rain was mixed with snow while today the sun shown brightly on the men.

"It's supposed to get into the fifties," noted one man. "Stop by and see how we feel next Sunday," another said.

One truck which arrived at the plant this morning was parked along side the street. The driver went to the office to notify the management that it was there and that he wouldn't drive it through the picket lines.

Another was parked at the front of the old plant. It was from Minnesota and was originally sent to pick up some farm machinery.

On June 7, 1973, Gehl Company and Local 579 negotiators shook hands on an agreement that the union would take to the membership. The average wage gain was eighty-five cents per hour over three years. The membership ratified the new contract 367 to 261.[185]

The strike ended, members of the local went back to work, and business returned to normal. The full ramifications of the labor dispute would not manifest themselves for years, but one decision at the time had an immediate effect. Gehl had been preparing to set up an assembly line at West Bend for the manufacture of skid-steer loaders. When Local 579 walked out, the Company revisited those plans.

In mid-May, Gehl announced its decision to build the new model HL4400 skid-steer loader in a new factory located in an industrial park in Madison, South Dakota. Located about fifty miles north and west of Sioux Falls, the initial building had been built by the Madison Development Corporation on a speculative basis.[186] "The move to another part of the country will give Gehl production and market flexibility, especially on new products," Joe Zadra told reporters.[187] For the next thirty-five years and beyond, Gehl Company would be a tenant of the South Dakota industrial park.

A Gehl model HL4400 skid-steer loader, painted by Dennis Crass of West Bend for Gehl Company's bicentennial "Showdown" program held in Las Vegas in 1976.

Minnesota. Carl Gehl and Joe Ecker negotiated an agreement to put Gehl Company's name on it for sale into the agricultural market.[174]

Several years later, Gehl Company designed its own model and started having it manufactured at the Hydra-Mac plant in Thief River Falls. It was an 800-pound machine that was much smaller than the Ericsson machine. The new unit broadened Gehl Company's agricultural product base because it was used for cleaning out barns.

Gehl, however, quickly concluded that it would never control its new product line until it started manufacturing the skid-steer loaders itself. In the summer of 1973, it began plans to build its own line of skid-steer loaders in Company-owned facilities.

Round Balers

Another 1970s diversification effort involved Gehl Company's decision to begin manufacturing round balers under license from another manufacturer.

In the early 1950s, New Holland introduced the first successful, automatic pickup, self-tying square baler. Over the next twenty years, square bales displaced the haystacks that had been a signature of American agriculture since colonial days. In the 1970s, manufacturers began introducing equipment that made large round bales of up to six feet in diameter. The round bales contained as much as a ton of hay, equivalent to up to forty or fifty square bales.[175]

The acknowledged leader in round baler manufacture was Gary Vermeer of Vermeer Company in Pella, Iowa. Founded fifteen years earlier to manufacture a mechanical hoist to unload farm wagons, Vermeer had brought out a large round baler in 1972.[176] Large round bales of hay and straw were handled mechanically and their storage resulted in less spoilage.

"We were the first licensee of the Vermeer Company," explained Joe Ecker.[177] Vermeer sold Gehl the right to manufacture and sell round balers. Gehl paid the Iowa company a flat license fee for the right to manufacture the round balers. Farmers loved the large round bales because they were much easier to handle. "The small square bale weighed forty pounds and was hard work to handle," Ecker explained. "You had to handle it by hand, and when you stacked them for storage, there was a lot of spoilage."[178]

In the mid-1970s, Gehl Company was still focused on gaining a larger share of the industry-wide sales of approximately 20,000 forage harvesters each year. Despite a strike, a sixteen-month recession, and nationally mandated price controls, the Company's operations weren't severely affected as sales doubled and gross margins increased 30 percent. By 1974, industry and Gehl Company forage harvester sales were at their maximum; however, each year thereafter experienced a decline in sales. What the Company didn't realize at the time was that the forage harvester market would shrink dramatically during the 1980s.

The Gehlbale 1500, the first round baler manufactured by Gehl Company, was launched in 1974. The designation "Gehlbale" was chosen to help people correctly pronounce the name of the Company.

GEHL 3000

Chapter Six

HARD *Times*

1977–1984

In the late 1970s, hard times returned with a vengeance for Gehl and the agricultural sector of the U.S. economy. Commodity prices began a sustained tumble in 1976 and 1977, caused in part by a rapid run-up in energy prices that had begun with the Arab Oil Embargo in the wake of the 1973 Yom Kippur War in the Middle East. In 1974 and 1975, prices at the pump doubled and tripled and doubled again.

But the energy crisis was even more pronounced for the nation's farmers. Tractors and agricultural implements gulped gasoline and diesel fuel, and much of the fertilizer that farmers spread on their fields was petroleum-based. And agricultural commodities moved to market in trucks, railcars, barges, and ships, all of which consumed huge amounts of fuel.

Inflation was an insidious affliction, affecting farmers more than energy shortages. President Lyndon Johnson's attempt to finance the Vietnam War in the late 1960s with a peacetime economy caught up with the nation in the mid-1970s. Inflation crept up throughout the 1970s, and by 1977 and 1978, inflation had risen above 10 percent per year.[188] Workers and unions were on strike almost constantly during the late 1970s in an attempt to recover the buying power that inflation had stolen from their wages. Farmers, unfortunately, could not go on strike to recover lost income.

After an outstanding decade of the 1970s with sales growth tripling and nearing a coveted $100 million in sales, things came to a head in 1979 when the nation's farm sector was hit with a triple whammy. Paul Volcker, who became the twelfth chair of the Federal Reserve in August 1979, realized that inflation had to be curbed. Accordingly, in October 1979, he initiated a series of interest rate increases that topped out at 19 percent in 1981. At the same time, the nation entered a deep recession characterized by unemployment rates that spiraled upward to near 10 percent by 1982.[189]

Economists dubbed the resulting economic condition "stagflation," which was particularly cruel for the agricultural sector of the economy. The nation's farmers had enjoyed boom times in the 1970s as land values soared and export prices buoyed domestic markets. Farmers were also able to take advantage of improved agricultural implements and machinery, wider use of hybrid seed, new fungicides and pesticides, chemical fertilizers, and close-row planting techniques.[190] But the stagflation of the early 1980s found many U.S. farmers financially overextended, especially those who had borrowed to purchase additional land in the late 1970s. The fall of commodity prices after 1980, spurred by President Carter's embargo of grain shipments to the Soviet

The Company closed the 1970s with the introduction of several new models of skid-steer loaders. Among those launched, the model HL2600 made its debut in 1977.

Union in the wake of the 1979 invasion of Afghanistan, was a particularly cruel blow to the nation's farmers. Equally cruel to the nation's farmers was the cut-off of Iranian oil to the United States in the wake of the 1979 overthrow of the Shah of Iran and the storming and takeover of the American embassy.

The early 1980s were terrible years for the U.S. farm sector. And when the farm economy caught a cold, Gehl Company caught pneumonia.

Plunging Profits

The farm recession with its low commodity prices rippled back into the offices of Gehl Company management in West Bend. In 1979, John Kamps was elevated from executive vice president to president. Joe Zadra became chairman and CEO, succeeding Albert Gehl, who had retired from the Board in 1975. For the new management team, introduction to the agricultural industry economics of that time was a shock. Interest rates were as high as 16 percent. Long-term rates were at 10 percent, and inflation was very high. "We incurred our first loss here in 1982," Zadra said. "In the early 1980s, there were lots of mergers and acquisitions in the industry. Gehl was financially sound, so we did not incur a lot of debt. The $10-million loan that we took out in the early 1980s was the largest debt that we had ever incurred."[191]

While 1981 had been a profitable year for Gehl Company, the problems that continued to plague the farm economy finally began to affect the Company's operations. Higher interest rates, high operating costs, and low commodity prices whipsawed farmers. With net farm income continuing to decline, the demand for farm equipment weakened to lower and lower levels. Such overall declines in farm productivity required less industry manufacturing capacity to fill the farmers' reduced needs.

Net revenues of $49,160,000 for 1982 showed a remarkable decrease from 1981's total of $74,634,000. The Company also went from a net profit of $1,466,000 in 1981 to a net loss of $2,110,000 in 1982. This was the first loss the Company had suffered since it was reorganized in 1934 and 1935. The next two years showed slight increases in sales through 1984, but in 1985, net revenue decreased to $38,561,000, which resulted in a net loss of $500,000.

Joe Zadra recalled that all agricultural implement manufacturers over-

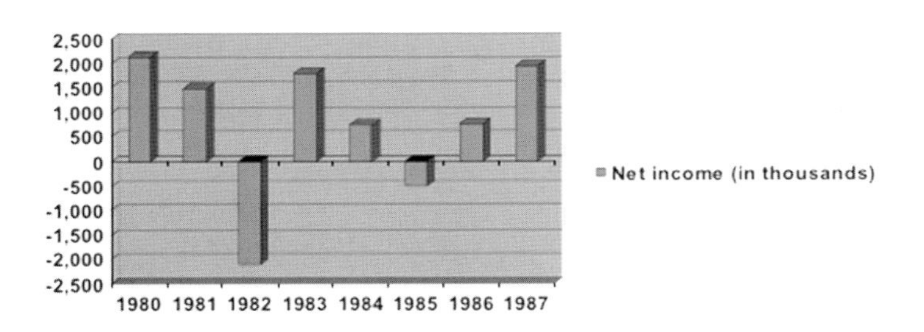

1980–1987 Net Income.

Joseph J. Zadra served as president from 1972 to 1979 and again from 1983 to 1985; chief executive officer from 1979 to 1985; and chairman of the Board from 1979 to 1994.

W.B. News 3-18-82

Gehl to lay off 35; county job outlook dims

By TOM LYONS
News Staff Writer

The Gehl Co. will lay off 35 employees at the end of their shifts Friday, as Washington County's already grim unemployment situation deteriorates.

A Gehl spokesman today said the layoff brings to 144 the number of people waiting to be called back to work. The layoffs were the result of "a readjustment of production schedules" necessitated by "continuing adverse business conditions," he said.

Cal Langer, Washington County Job Service supervisor, said the layoffs at Gehl come on the heels of recent layoffs at Sherwood-Medical in Jackson and Regal Ware in Kewaskum.

Regal Ware last month laid off approximately 110 workers when other employees came back from voluntary layoffs. At Sherwood-Medical,

(See GEHL page 14)

• Gehl

(Continued from page 1)

formerly Searle, 34 of 135 plant personnel were laid off last week, a company official confirmed this morning.

The Sherwood spokesman also announced that company is in the process of being sold by Brunswick, its parent corporation, to American Home Products, a New-York based conglomerate of pharmaceutical and consumer products. The sale, which is now being finalized, will have no effect on the local employment picture, the spokesman said.

Langer said his office has processed about 125 new applications for unemployment compensation on each of the past three Mondays. The lines of jobless workers will grow longer, he predicted.

The January unemployment rate was a record-high 10.6 percent, Langer said. "My gut feeling is that the figures for February (not compiled yet) must be over 11 percent. I wouldn't be surprised if it went over 12 by the end of March," he said.

In 1981, by comparison, the county's unemployment rate was 9.7 percent in February and decreased to 9.0 in March, Langer said.

Gehl to lay off 65 at shift's end

By TOM LYONS
News Staff Writer

WEST BEND — The Gehl Co. will lay off another 65 workers at the end of the second shift today, bringing to 191 the number of employees laid off since April.

The layoff reaches back to 1977 in the seniority rolls of Local 579 of the Allied Industrial Workers, said Art Lastofka, company vice president of industrial relations.

Gehl, Washington County's second largest employer, laid off 100 of its 739 union-represented employees in late April. Another 26 were laid off in May.

The company, which manufactures farm equipment, is currently suffering from the economic crunch that has kept farmers away from large-ticket purchases. Gehl employs about 1,200 people locally and an additional 200 at its South Dakota plant. Both plants have laid off approximately equal percentages of employees, Lastofka said.

While layoffs at the West Bend headquarters have been restricted to union-represented workers, Lastofka said, the non-union people are also paying their dues.

Between 15 and 20 clerical workers have voluntarily cut back their hours and their pay, Lastofka said, forestalling the necessity of layoffs among the non-union people.

Starting in late June, he said, some clerical workers began to work four-day weeks or to take several days off without pay during pay periods. "We took this approach because people may like the time off, particularly in the summer months," he said. "We've been very flexible."

While the voluntary cut in hours has prevented non-union layoffs, such cutbacks are apparently still possible. More union layoffs also loom as a possibility, he indicated.

At this point, said Lastofka, Gehl's does not see the light at the end of the recession tunnel.

The situation at the county's largest employer is brighter.

The West Bend Co., which normally employs roughly 2,300 people, is holding its own, according to a company spokesman.

Currently 64 employees are laid off, said Larry Grescoviak of the industrial relations department. However,

(Continued on page 12, col. 3)

• Gehl

(Continued from Page 1)

natural attrition has reduced the total work force number to about 2,000, he added.

"At the present time, we are planning on simply maintaining the status quo," said Grescoviak. "We're just holding steady."

At Amity Leather Co., another major employer in the area, the layoff situation bottomed out earlier this year and has now started to bounce back.

Amity, which employs roughly 500 factory workers, has laid off approximately 170 workers this year, according to a company spokesman.

But recently, the company began calling some of them back. Currently there are about 100 Amity employees who are still laid off, the spokesman said.

The overall unemployment situation in the county seems to have levelled off for now, according to Cal Langor, office supervisor for Job Service.

"It's holding — probably slowed down a little," said Langor. "There's still going to be some more coming from what I hear, but the pace has slowed down."

The June unemployment figures place Washington County's unemployment at 7.9 percent, Langor said, slightly ahead of the state figure of 7.2 percent and the national percentage of 7.8.

But Langor said the county figure reflects a large number of people who work in nearby Ozaukee and Dodge counties. "That has to be a lot of it," he said. "Most of the biggies in the county are holding up good."

The 1980s began as a challenge to Gehl. With three successive farm recession years beginning in 1980, the Company had no choice but to lay off workers and shut down plants on several occasions.

Gehl to close 6 more weeks this year

9-21-82

4-week shutdown ends

By TOM LYONS
News Staff Writer

Hundreds of Gehl Co. workers returned to work Monday after a four-week shutdown of the plant to news that the plant would close again for six weeks at the end of the year.

Gehl President John Kamps notified returning employees Monday that a previously-announced shutdown of two weeks at the end of the year had been changed to a six-week shutdown. The plant will close from Nov. 22 to Dec. 31.

Orders for farm machinery have not picked up, Kamp said, and the shutdown relfects a backlog of inventory.

"I've been around here 31 years, and this is the first year we've ever shut down our facilities," he said.

The Gehl Co., one of West Bend's major employers, shut down for four weeks beginning in April and for another four-week period that ended Monday. The end of the year shutdown will bring total plant closings to 16 weeks.

Gehl's plant in South Dakota shut down in April. It did not close in September and will close for only two weeks at the end of the year, the president said.

The West Bend plant closes and then calls everyone back, Kamps said. The South Dakota plant closes less often and for shorter periods, but operates at a reduced level compared to West Bend.

The West Bend plant currently employs about 450 plant and office workers, according to Art Lastofka, Gehl vice president of personnel. The company has laid off hundreds of workers in the past two years.

The recession that has crippled the farm machinery industry for the past two years has cut the West Bend production staff at Gehl from a high of 750 in early 1980 to 264, said Lastofka.

Kamps said the woes of Gehl and the farm machinery industry as a whole were tied to difficulty farmers were having scraping together the money to buy machinery with the high cost of money and the low price on crops.

"In a good many product (lines), the industry is selling at less than 50 percent of peak sales (years), " Kamps said.

Gehl's sales peaked in 1979 at about $100 million, Kamps said.

Despite recent lowering of interest rates, Kamps said he was skeptical about any major improvement in Gehl's situation in 1983.

"To get better, we need improved crop and livestock prices, lower financing costs and (better) availability of financing," Kamps said.

"And somewhere we need a different attitude on the parts of some farmers and people," the company president said. "There's a lack of confidence that sort of permeates all of our lives these days."

The farmers' crunch has begun to hit the pocket books of Gehl investors this year.

Despite a poor sales year, Gehl managed to pay dividends to stockholders for all four quarters of 1981 and the first quarter of 1982. The past two quarters the company has not paid dividends.

But Kamps said he was unwilling to write off 1982 as money loser.

Kamps also said the company was strong enough to ride out even a prolonged, industry-wide recession such as the one in which it is currently mired.

"We have taken the steps we have to date (the shutdowns) in an effort to survive this difficult period, which has not ended," he said.

"But we are a financially strong company and we intend to conserve our resources, so that when business improves, we will be in a strong, competitive position to move forward," Kamps said.

In a news release distributed to employees Monday, Kamps said that even during the year-end shutdown, the "shipping of machinery and service parts will continue ... with minimum levels of employees maintained."

Lastofka said it was not yet determined how many employees would work during the shutdown.

Reduced demand throughout the early 1980s resulted in a great deal of Gehl equipment remaining in storage lots, as shown here in 1983, as the farm economy continued to struggle. *(Photo courtesy of* The West Bend Daily News*)*

produced during the 1980s. To counteract these negative forces and to help bring inventories into line with sales, Gehl took the dramatic step of closing its West Bend plant for a total of thirteen weeks during 1982. The Company also closed its Madison, South Dakota plant for a total of six weeks that same year.

Gary Rentz, who retired in 2008 after serving as the Madison plant's human resources manager since 1991, started to work for Gehl Company in April 1979 at the firm's Madison facility. The native of nearby Wentworth, South Dakota, started on the assembly line on second shift, and within six months was promoted to supervisor on the first shift. "Gehl had quadrupled the size of the Madison plant during the 1970s, and in 1979, the facility had a workforce of more than 275 people," Rentz explained.[192] Four years later, however, there were only fifty production employees on the line at Madison.[193]

By the beginning of 1983, the need to balance factory and field inventories with retail sales became the main focus of the Company. Because the previous year's plant closings had not been entirely successful, the Company took the step of keeping the West Bend plant closed for the first six weeks of 1983.

While cost containment had always been an objective in the operation of the Company, it received particular attention during this difficult five-year period. In addition to the temporary plant closings, Gehl reduced its workforce in 1982 and 1983 from 979 to 588 employees, primarily through layoffs. Shortened workweeks during this period also were instrumental in reducing labor costs.

On the Rebound

For Joe Zadra, the slowdown of the early 1980s was a difficult environment in which to manage. Zadra, who relinquished the Gehl Company presidency in 1985 to move up to chairman of the Board, recalled that "during the 1980s, the industry's larger companies had expensive incentive programs and retail finance programs. We developed our own retail finance company. The terms to the dealer were very long. We often sent inventory to our dealers, interest-free, for as long as twelve to eighteen months. From 1981 to 1985, our sales dropped from $74 million to $38 million."[194]

The farm landscape changed significantly. "We had 20 to 25 percent of the market, but in the late 1980s and early 1990s, the market just disappeared. We allied ourselves with dealers who didn't have a full line of implements, and this had a very adverse effect on the Company."[195]

Gehl Company had a short stint manufacturing self-propelled forage harvesters. The self-propelled Gehl 188 Chop-King is shown here with the hay pick-up attachment. Production of the machine ended in 1969.

{ GEHL FINANCE, 1984 }

One of the factors that contributed to the remarkable turnaround at Gehl Company during the late 1980s was its continuing strong relationships with its dealers. Since the late 1940s, Gehl dealers had been the Company's strength. The Gehl sales force fanned out across the country, often with 16-millimeter film projectors in the trunks of their cars. Dealers traveled to West Bend once or twice a year to preview the next year's model of forage harvesters and other equipment, and Gehl kept dealers informed about the Company and its products in the popular Company periodical, *Choptalk*.

"In the 1970s, our sales people got lazy," explained Terry Lefever. "They were just writing orders and not selling anymore. The early 1980s were a tough time for the Company. We had to ask the question: Are people buying Gehl products or are they buying products from Gehl people?"[213]

By 1980, Gehl had almost 900 dealers, and the average annual dealer volume was $80,000. The Company realized that 20 percent of its dealers were doing 80 percent of the business. But all the dealers wanted the profitable Gehl parts business. Gehl had to figure out how to drop the smaller dealers but still retain them as friends.

"We were going through a review process, but upon completion, we were just putting the reports in a folder and filing them away," Lefever said.[214] A set of standards for the ideal Gehl dealer was put together, and the territories with the best potential were selected for retention.[215]

With that, Gehl started to collect baseline data that allowed it to start rating dealerships. The Company gave dealers bonuses based on their ratings and gave them a one-year contract for parts at a discount. In just a few years' time, Gehl reduced the dealer list from 900 to 450 but doubled the average annual dealer volume. The restructuring created some dealers with a sales volume of $1 million or more per year.[216]

With the number of dealers pared, Gehl took another step that ensured the success of the stronger dealerships. In 1984, the Company established Gehl Finance, a division dedicated to providing retail financing to farmers interested in purchasing Gehl equipment. Gehl had provided financing to dealers since shortly after World War II, and dealerships were allowed to return unsold inventory. But retail financing was new for Gehl Company.

Gehl and other agricultural equipment manufacturers had identified retail financing as a key element in marketing and selling products. Shortly after the program was rolled out in 1984, nearly 90 percent of eligible Gehl dealers elected to participate in Gehl Finance.[217] Dealers liked the program because they could book profits when the equipment was sold to the farmer.

Within a year, Gehl Finance was operating in eighteen states, including all of Gehl Company's large-volume states. By the end of 1985, Gehl Finance's outstanding contract receivables totaled more than $2 million, a ten-fold increase from the end of 1984.[218] Five years later, at the end of 1989, the Company was servicing $52 million of retail financing contracts, a measure of the key role Gehl Finance played in the Company's resurgence during the latter half of the 1980s.[219]

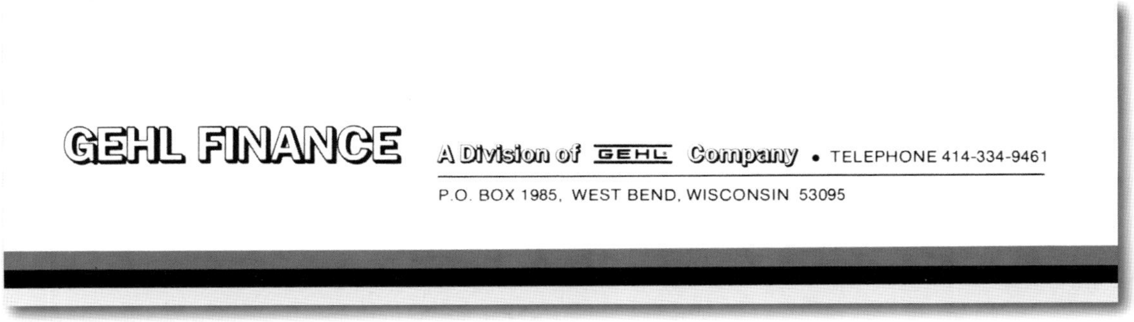

Joe Ecker recalled that the Company's financing arm created problems, even in the good years. "We had a financing arm that financed equipment to the dealers," Ecker said. "Sometimes we would have accounts receivable totaling $100 million or more. Almost every state had buy-back laws that allowed the dealers to return unsold inventory, so we had to train salesmen in managing the dealers' inventory. We got back a lot of weathered equipment, so it behooved us to control that dealer's inventory. What you would do, invariably, was try to transfer it and discount it. You would try to keep that to a minimum, but it was still a fact of life. Some of the state laws also contained a provision for a buy-back of parts, and that could really be a problem."[196]

As the Company moved into the mid-decade, the introduction of new products and a strong budget austerity program brought a return to profitability, although 1985 would prove to be the low point of the decade in terms of sales and operating losses. Gehl had positioned itself to benefit from the beginning of an economic upturn. Sales rebounded spectacularly by 1990, reaching a level of $175,000,000, which resulted in a net income of $7,318,000.

Gehl Company's recovery from the difficulties of the early 1980s resulted from a concurrence of unrelated actions. The Company embraced computerization, it brought out new agricultural implement product lines, and it diversified into construction equipment.

The Computer Revolution

Bringing automation to the Accounting Department is always traumatic for employees used to doing everything manually. Manual typewriters, carbon paper, hand-cranked adding machines, and mimeograph machines gradually were replaced by personal computers, copiers, and fax machines. Employees sometimes grew quite anxious when they saw how the new technologies would affect their jobs.

Laura Heisler, who worked for Gehl for fifty-five years, remembered her early career in the Company Payroll Department. "I started in the Payroll Department in 1948, where I would keypunch factory cards," she said. "I ended up doing all the shop and office payrolls. At one time, I had 750 people in the shop,

Inside the Gehl office in the late 1970s.

Gehl Company's office kept pace with the transformation to computer technology that occurred in the 1980s. In fact, Gehl was one of the pioneers in the application of on-line computer-based systems and implemented a company-wide integrated decision support system, which allowed individuals from all plants to access the same information. This was especially important with the Company's numerous plant locations.

and I did the payroll all by hand. I calculated hours and payroll taxes all by hand."[197]

Joe Zadra recalled the switch to office automation in the early 1980s. "With the onset of computerization, we hired an individual from A. O. Smith, who had experience with automated equipment in manufacturing processes," he said. "In the very early 1980s, we installed an IBM System 360 and used it for payroll and sales. We later integrated the manufacturing processes."[198]

Product development experienced significant advancement with the computer revolution. The Test Department was able to simulate various applications and evaluate machine performance both in the laboratory and out in the field.

The continuing upgrade of the Company's Accounting Department to enable it to support the long-term growth of the Company was augmented with the launching of a five-year data processing plan in 1985. The plan was headed by Dick Semler, vice president of data systems and administration, and was known as the Data Processing Plan for Systems, Hardware/Software, and Office Accounting. It was established to enable the various divisions of the Company to support reduction of operating costs and to evaluate new business acquisitions. At that time, the Accounting Department was provided with the first generation of personal computers when three PCs were provided to employees in the corporate office.[199]

With the onset of additional cutting-edge technologies such as Just-In-Time inventory management and Total Quality Control, Gehl management had positioned the Company to more closely monitor production, external growth, and product quality. The initial installation of the IBM System 360 computer provided the platform for Gehl to integrate its accounting and production systems and develop more sophisticated methods of financial management. And years later, with the passage of the Sarbanes-Oxley Act in 2002, which required enhanced reporting standards for publicly held companies, Gehl would be able to meet the compliance standards because of the integrity of its financial controls.

Computerization allowed Gehl to become far more productive across a range of departments and job functions. "In 1984," said Zadra, "we installed CAD/CAM software in our Engineering Department, and this really shortened the time to develop new products. We had 13,000 components and assemblies in our inventory. We also installed the Just-In-Time purchasing program. We bought our steel from the service centers when steel prices went sky-high in the mid-1980s. Gehl was noted for its field service assistance. We had a well-trained service force out in the field. We sought people who were mechanically oriented and grew up on a farm over college graduates who weren't familiar with farming."[200]

One notable exception to this selection criteria was Hughel ("Gene") Miller. Dick Gehl would often make early-morning stops at the local butcher shop and was quick to notice one of the butcher shop's employees, Gene Miller. Gene's winning personality and the obvious fact that he was an early riser made him a desirable employee in the eyes of Dick Gehl. It wasn't long before Dick Gehl recruited Gene to work at Gehl Brothers, where Gene remained for many years and eventually became a field sales manager, traveling the world representing Gehl products.

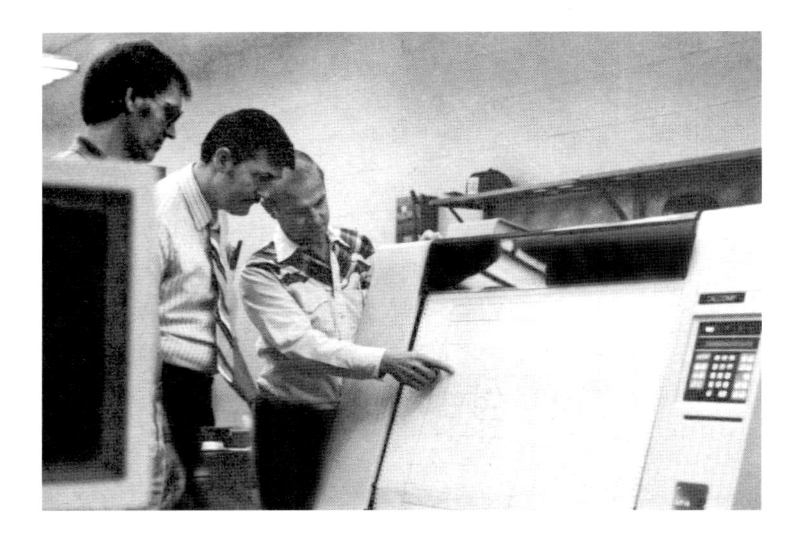

The engineering group, represented here by (from left to right) Chuck Kaehler, Jim Shutes, and Dick Junge, quickly discovered that improved productivity and reduced time in product design were just a few of the benefits of the Lockheed/IBM CADAM system the Company installed in mid-1984.

The Gehl model CB1200 heavy-duty forage harvester was designed to match up with high-horsepower tractors, with a drive train built to handle an input of up to 225 horsepower.

New Product Lines

Gehl Company began a decade-long upgrade of its product lines in 1974. Joe Zadra recalled that "between 1974 and 1979, we came out with new model forage harvesters, which had larger capacity. There were seven manufacturers at that time producing forage harvesters, including such names as Deere, Massey-Ferguson, Case, and New Holland. By now, farms were getting larger, and our competitors had cylinder-type forage harvesters while we had flywheel forage harvesters. In 1968, we built the East plant, installing the Harrick Forager, which was a high-lift crane that allowed one man to handle the steel. In the old West plant, it took six to eight employees to handle the steel."[201]

As farm size continued to grow, tractor horsepower became increasingly higher. To respond to this need, the trailing equipment needed to be upgraded to match these larger tractors. In 1977, the forage harvester line was updated to span from the 120-horsepower limit of the model CB700 to the 225-horsepower limit of the model CB1200.

The 1977 introduction of the model CB1200 boasted a first-of-its-kind innovation: the model CB1200 had an offset blower instead of a blower mounted behind the cylinder, which helped contribute to improved efficiency. This forage harvester, after development testing, was put into full production at the West Bend plant in 1978.

In January 1981, Gehl announced a new generation of round balers, designated as models RB1450 and RB1850, which embodied the newly patented "Total Density Control" air-hydraulics system. With adjustable air pressure, the operator

The Gehl model BH1500 bale handler, as shown here, and the bale carrier entered the Gehl product line in 1976, but were discontinued just five years later.

GehlFest '78 was a special dealer meeting held in July 1978 to familiarize Gehl dealers with the Company's new product offerings and facilities. All Gehl dealers in the United States and Canada were invited to attend, and of those dealers, more than 2,600 representatives attended. Here, Vince Leach promotes the newly introduced Gehl model CB700 forage harvester.

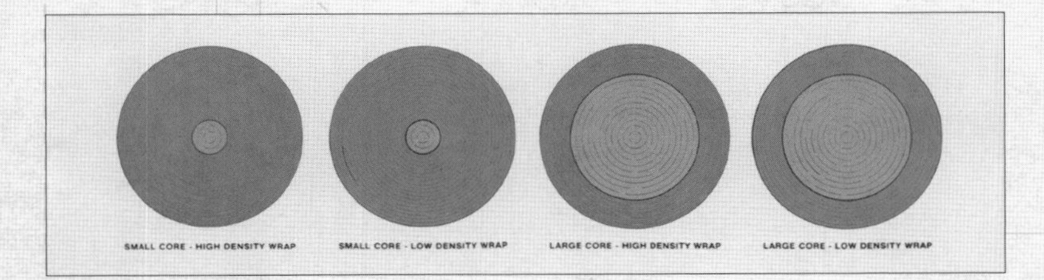

The four baling densities produced by Total Density Control.

The model RB1850 round baler, featuring Total Density Control, was able to make bales literally hundreds of pounds heavier than conventional round balers.

of the baler could control the size and density of the bale's center core. Then, with the activation of the hydraulics, the outer bale wrap began and the operator could control the density, from loose to dense.

In a short haying season, this density control feature allowed operators to bale their crop in less-than-ideal conditions and still produce top-quality bales that stood up to weather and retained crop nutrients.[202]

Because moisture levels in the crop govern bale density, this feature allowed farmers to set the way each bale was made. The operator could make bales with dense center cores and dense wraps; soft center cores, up to three feet in diameter, with a dense wrap; or loosely wrapped bales with a loose center core.[203] Bales made under optimum conditions could be hundreds of pounds heavier than conventional round bales.

The model RB1450 baler made bales that were forty-five inches wide and up to sixty inches in diameter; the model RB1850 baler made bales that were sixty inches wide and up to seventy-two inches in diameter.[204]

In 1982, the water tank option was introduced as an attachment to Gehl Company's line of forage harvesters. The water tank option was one of the more popular innovations because it reduced haylage gumming and the power requirements for harvesting crops. The CB750 forage harvester was updated with many of the features already available on the larger models CB1000 and CB1250.

A New Harvester Line

In 1984, Gehl introduced the totally new, state-of-the-art 60-Series forage harvesters. The CB760, CB1060, and CB1260 models improved capacity and efficiency while remaining affordable to the dairy and hay farmer. The CB1060 and CB1260 featured Gehl Company's AutoMAX load sensing and maximum capacity design. Load sensors on the units not only automatically shut down the harvester in the event of an overload, but also allowed the operator to electronically re-engage the clutch and process material without having to leave the seat.[205]

Rated for tractors up to 300 horsepower, the CB1260 was the biggest forage harvester Gehl had ever produced. Both the

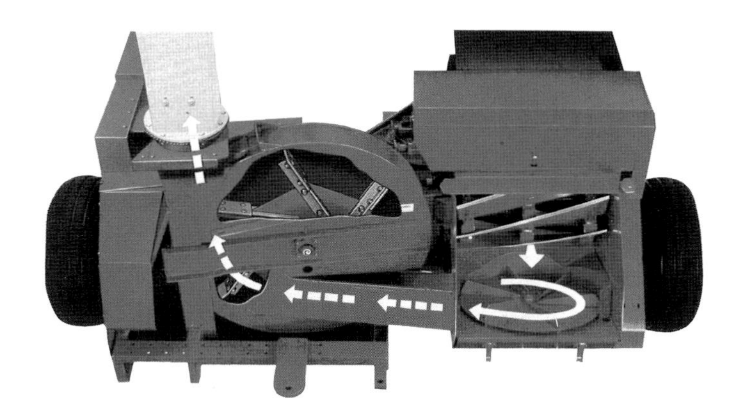

A new concept in forage harvesting: the AutoMAX system, unique to the model CB1260 and CB1060 forage harvesters, was designed to protect against overloads and increased material-handling times.

As Gehl Company entered the compact equipment field with its skid-steer loaders in the mid- to late 1970s, it targeted a variety of customers, such as farmers, contractors, and municipalities, by offering a variety of attachment options, compact size, and affordability.

In the midst of a struggling farm economy, Gehl introduced its largest-capacity forage harvester yet, the model CB1260.

Farmers using harvesters with the Metal Stop system were likely to pick up an assortment of materials that were prevented from entering the cylinder.

CB1260 and CB1060 were capable of being equipped with a Metal Stop ferrous metal detector that protected the harvester from unwanted tramp metal entering and damaging the unit's cutters and processing machinery, which in turn protected livestock from injury due to ingesting pieces of metal.

Rounding out the new introductions in the early 1980s were the FB1540 and FB1580 forage blowers, and the DC2330 and DC2350 disc mower conditioners. "The addition of these units makes the Gehl hay-tools line one of the largest available today," the Company proclaimed in 1984.[206]

Even with the introduction of strong new product lines, Gehl, like the rest of the industry, struggled to remain profitable. In 1984, Gehl initiated an eleven-week plant shutdown at West Bend in an attempt to balance inventories and sales.[207] That came on top of a four-week shutdown in the late summer of 1982 and a twelve-week shutdown in the winter of 1982–1983.[208]

Changes At the Top

The depressed conditions in the agricultural equipment industry took their toll on the Gehl executive management team. In 1983, John G. Kamps, who had been president of Gehl Company since 1979, announced his resignation. Kamps had joined the Company in 1951 as an accountant. He was subsequently named assistant credit manager in 1953; credit manager and assistant treasurer in 1963; treasurer and assistant secretary in 1969; and executive vice president in 1972.

John G. Kamps, president and chief operating officer from 1979 to 1983.

When Kamps took over the operation of the Company as president, Gehl was experiencing a period of steady growth, with sales peaking that year at $100 million. But that period of prosperity suddenly ended in 1980 when the recession crippled the economic health of the farming industry. Kamps was forced to preside over several temporary plant closings in 1982 and early 1983 to help bring inventories in line with sales. Company employment also plummeted from a high of 1,477 in 1979 to 908 by March 1983.[209]

Reflecting on those tumultuous times in the agricultural industry, Kamps noted that "it was the longest and most difficult situation I had seen in my thirty-one years in the business."[210] It would be three more years before sales would begin to rebound and another year after that before sales would surpass the levels achieved in 1981.

On December 12, 1983, Kamps abruptly resigned as president and chief operating officer of Gehl Company, citing personal reasons and a desire to take some time off. Joe Zadra, chairman and chief executive officer, said at the time that "Kamps' resignation came as a complete surprise to us."[211]

On October 1, 1984, Kamps accepted the position of president of Kasten, Inc. of Allenton, Wisconsin. At the time of his acceptance, he remarked that he took some time off after resigning from Gehl, "but I always intended to find another position. I wanted to do something."[212]

Kamps' resignation opened the way for new leadership at Gehl. And that new leadership would preside over a stunning reversal of the Company's fortunes during the second half of the 1980s.

GEHL

Chapter Seven

Transformation

1985–1989

For Gehl Company, the late 1980s was a transforming era in the Company's long history. In the relatively short span of less than five years, Gehl broadened its product line through a series of acquisitions, diversified into the construction market, and expanded into international sales. To cap its late-1980s transformation, Gehl became a publicly held stock company in 1989.

Mike Mulcahy, who joined the Company in 1974 as corporate counsel, recalled that during his first ten years of employment, Gehl was unchanged from what it had been for decades. "Gehl was an old-line family-held Wisconsin company," he said. "It was privately held with an excellent reputation. It was a heavy-equipment manufacturer with two plants in West Bend and a newly opened facility in Madison, South Dakota. The Company had just come off a protracted thirteen-week strike with the AIW. They had had a good relationship with the union, but there was a lot of bitterness after that strike. The Company never did shut down."[220]

Previously a family-dominated board, the Gehl Company Board of Directors was transformed by Joe Zadra in the mid-1980s from a board composed of insiders to one dominated by outside directors.

Different Strategies

"In the period from 1985 to 1988, the Company started looking at different strategies," Zadra explained. "We wanted someone from the outside to be president, someone who was aggressive and from the construction industry. In 1987, we had a resurgence in sales, climbing to $85 million with a $2-million profit. This was all due to the skid-steer loader and the round baler."[221]

The Company's strategic direction for agricultural equipment manufacture in the latter half of the 1980s was governed by a simple equation. Some 2,000 U.S. agricultural equipment dealerships had closed between 1980 and 1985, and the competition in the tractor and self-propelled combine sectors of the market was fierce. Gehl Company's strategy was to position itself as the national manufacturer and distributor of specialized agricultural equipment that logically complemented a dealership's major tractor line.[222]

The strategy for the agricultural equipment segment of the Company's business further envisioned a broadened product line to serve its targeted agricultural customers, typically livestock, dairy, and poultry farmers. Gehl resolved to dominate five product

By offering complete product lines within those areas in which the Company specialized, such as haymaking, Gehl was able to increase its market share in those segments of the agricultural equipment industry it served.

lines in that marketplace, including haymaking, forage harvesting, feed making, material handling, and manure handling equipment.[223]

The Gehl strategy would be carried out by new executive leadership. It would involve, for the first time in the Company's history, acquisition of other manufacturers. When John Kamps left Gehl in 1985, the Board began a nationwide search for a new president and CEO. They found their candidate in Bernard Nielsen, a veteran of agricultural equipment manufacture and general manager of Coldwater, Ohio-based New Idea Farm Equipment Company, a manufacturer of hay and forage equipment. A graduate of the University of Michigan, Nielsen's professional experience included employment with Fisher Controls and HawkBuilt Company. He had come to know Gehl management through his service on the Board of Directors of the Farm and Industrial Equipment Institute.[224]

Nielsen immediately put Gehl on an acquisition course. By the late 1980s, the stagflation of the earlier part of the decade had begun to abate, and investment grade credit became far easier to obtain. Interest rates dropped continually throughout the period, and rates stayed low throughout the 1990s.

"During the Nielsen era, we obtained new manufacturing expertise and completed a number of acquisitions," Mulcahy explained.[225]

"In 1987," Joe Zadra added, "we acquired certain assets of TCI Power Products of Yankton, South Dakota. They made motor graders and rough-terrain telescoping-boom forklifts." It wasn't necessarily a high-volume business, but in good economic times, it could be very profitable, Zadra explained.[226]

Merrick Monaghan, the Yankton plant's manager of engineering, joined Gehl Company in Yankton about six months after Gehl Company purchased certain product assets of TCI Power Products. The Terry, Montana native had recently graduated from the South Dakota School of Mines in Rapid City and was looking for a job. "There were about one hundred people working in Yankton when I was hired," Monaghan said. "One of my earliest memories is that around that time, production was increased to one machine a day, about 350 a year. That was a really big deal back then."[227]

The late 1980s purchases were mostly asset acquisitions, the first in the Company's long history. In 1985, however, Gehl bought Hedlund Manufacturing Company, Inc., which made a well-patented line of manure spreaders. Hedlund owned

The Company strategically sold Gehl products at dealerships that carried a major tractor line.

The Dynalift name has been known to the industry since 1974. With further development, the machine was transformed from a simple lift-and-reach machine (above) to a very versatile telescopic handler (below).

two plants: one in Boyceville, Wisconsin, and a second in Myerstown, Pennsylvania.[228] Hedlund had previously acquired Martin Manufacturing Company from Mervin Martin, the inventor of the "V-Tank" manure spreader, and subsequently formed the subsidiary Hedlund-Martin, Inc.

It was a very successful product for Gehl, Mulcahy recalled, "but we inherited more legal issues than most lawyers would see in their entire careers. We even had a patent malpractice action filed against the company's outside patent lawyer who had allegedly forged Mervin Martin's signature on the patent application. On a positive note, royalties resulting from the Hedlund-Martin addition helped the Company's bottom line in the early 1990s when we needed it the most."[229]

At that same time, Gehl Company bought certain assets of several other agricultural equipment companies. One successful acquisition was that of certain assets of Van Dale, Inc. in 1988, which brought aboard a $12-million-a-year mixer-feeder line. "We also acquired certain assets of Champ Corporation, a straight-mast forklift company located in California," explained Mulcahy. "It was a very rugged machine, but we never really were able to do much with it. Today, we're not making the Champ products. They just didn't fit our line."[230]

In 1991, Gehl acquired a line of asphalt paving equipment, including pavers, rollers, and road maintainers from Puckett Brothers Manufacturing Company, Inc., located outside of Atlanta. The company made small asphalt pavers for golf courses and the rail-to-trail market. In early 1994, Gehl closed the Georgia plant and moved the product line to Yankton. Gehl continues to manufacture the PowerBox asphalt pavers.

OMC

In 1986, Gehl acquired the agricultural product line assets of Owatonna Manufacturing Company (OMC). "They made a self-propelled windrower, a pull-behind mower conditioner, and some minor hay tools," Zadra explained.[231]

OMC was typical of the problems affecting the farm equipment manufacturing industry in the 1980s. OMC's background paralleled that of Gehl Company. It was family-owned, and it made similar products. Founded in the southern Minnesota community of Owatonna in 1865 to make grain drills, seeders, and butter churns, OMC diversified into making farm elevators and storage lift systems in the 1920s, and began building self-propelled windrowers after World War II. In the late 1960s, OMC began making skid-steer loaders with a planetary drive based upon the drive used in the company's line of self-propelled farm machinery. In 1972, the Minnesota firm came out with a line of hydrostatic drive skid-steer loaders.[232]

OMC fell on hard times in the mid-1980s in the wake of the contraction of the U.S. farm sector. In 1986, OMC filed a Chapter 11 bankruptcy petition. The company employed 400 people at its plant in Owatonna at the time.

The purchase of assets from Van Dale, Inc. assisted Gehl in designing its line of mixer-feeders, which included truck-mounted, trailer-mounted, and stationary units. The initial five models ranged in size from 210 to 500 cubic feet.

The Gehl model 1438 PowerBox paver.

From the purchase of Hedlund Manufacturing Company, Inc., which was at one time a leading manure-handling equipment manufacturer, Gehl began manufacturing V-tank-type manure spreaders designed for both semi-solid and high liquid-content manure handling.

{ RESTRUCTURING THE COMPANY, 1989 }

Gehl Company's emergence as a publicly held company in November 1989 provided the executive team with the opportunity to restructure the Company. Shortly after Bernard Nielsen returned from the NASDAQ trading floor in New York, Gehl announced the formation of separate organizations within the Company: the Agricultural Division and the Industrial Construction (IC) Group.

"Each division has full responsibility for its own manufacturing, marketing, and service," Nielsen announced to shareholders early in 1990, noting that the Company had expanded and strengthened its executive staff to take advantage of the new synergies afforded by the reorganization.[250]

With $114 million in sales in 1989, the Agricultural Division of Gehl continued to produce the Company's flagship line of equipment. Increased support to dealers resulted in a 31-percent increase in the average dealer's sales volume of Gehl products during the year. In just five years, the Company had tripled its agricultural product line, from thirty-three to one hundred models. The new general manager pledged to elevate the Agricultural Division into a world-class manufacturing powerhouse with the implementation of processes such as Just-In-Time (JIT), Total Quality Control (TQC), and Focused Factories.[251]

The IC Group continued its rapid growth in 1989 by introducing new products, expanding its dealerships in strategically targeted markets across the United States, and strengthening its distribution both nationwide and internationally. With $40 million in sales in 1989, the IC Group had average growth of more than 200 percent each year since its founding in 1984.[252]

Employment had gone from a low of 400 workers in 1985 to more than 1,000 employees by the time Gehl went public in 1989. The number of IC Group employees increased dramatically with the rapid ramp-up in sales.

The Gehl IC Group was developed in the late 1980s to provide more focus on marketing Gehl skid-steers into the construction marketplace.

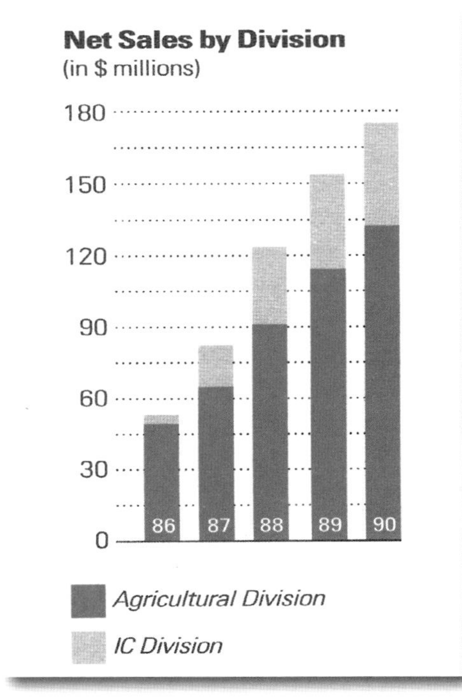

Net sales by division, by year. Industrial/construction equipment business accounted for 24 percent of sales in 1990; agricultural equipment sales were 76 percent.

At the time, the 1989 restructuring into two divisions, agricultural equipment and construction, made sense. Gehl was catering to two similar, but in many ways entirely different, marketplaces. And it was clear to executive management that light industrial and construction markets would continue to drive Gehl Company's growth overseas.

The IC Group logo.

One new product resulting from Gehl Company's asset purchase from OMC was the fixed-chamber round baler, which Gehl quickly introduced just one year after the line was purchased from OMC.

Gehl purchased the agricultural assets of OMC from the bank. "Gehl got all the agricultural equipment on the lot, parts, tooling, drawings, and specs," said Lyle Snider, a longtime veteran of OMC's Engineering Records Department.[233] "Gehl then took all the agricultural equipment manufacture to West Bend."

Several OMC employees moved to West Bend with Gehl following the purchase. Snider spent several weeks in Wisconsin helping the Gehl engineering staff sort out records and then returned to Owatonna to handle engineering records for OMC's remaining skid-steer loader line. But it wouldn't be the last time Snider would work with Gehl.

At the time of the bankruptcy, OMC was involved in protracted patent litigation with Bobcat. OMC's skid-steer loader line was sold from bankruptcy to Austoft Industries of Bundaberg, Australia, in 1987. Austoft, Inc. (U.S.A.) was its original corporate name. Austoft settled the litigation with Bobcat, changed the name of the company to Mustang Manufacturing Company, Inc., and continued producing the Mustang loaders at the plant in Owatonna. In 1989, BM Group, PLC of Chippenham, England (later known as Brunell Holdings, PLC), acquired Mustang and ran it for several years before selling it to Gehl in 1997. With that purchase, Gehl obtained the Owatonna manufacturing plant, and Snider finally wound up working for Gehl.

In addition to those mentioned, select product asset purchases of the 1980s also included assets from Keith Industries, Inc. and Kasten Manufacturing Corporation.

The Industrial Construction Group

Gehl faced many business cycles over the years. Mulcahy explained that "there have been three or four downturns since I came to the Company in 1974. One of the goals of the new Industrial Construction Group was to create counter-cyclical product lines to agricultural implements."[234]

Bernard Nielsen noted in 1985 that "the agricultural equipment market is relatively mature and has always been cyclical in nature. At present, and in the near future, it will continue to experience what I refer to as 'evolutionary times.'"[235] Nielsen went on to point out that "our commitment to agricultural equipment is resolute and undiminished."[236] But Nielsen, the management team, and the Board realized that the time when Gehl could afford to put all of its eggs in a single basket was long past. Gehl established its Industrial Construction (IC) Group in 1986 to meet the specialized needs of the light industrial and construction markets. By early 1988, the IC Group had achieved an important presence in virtually every major U.S. metropolitan market. At that time, Gehl had more than 125 IC Group distributors across the nation.[237]

Joe Zadra recalled the 1986 decision to diversify as one that seemed utterly natural at the time. "We developed the Industrial Construction Group to help the future growth of the Company," he said. Nielsen brought in additional officers to staff the reorganized Company.

As part of its five-year strategic plan, Gehl dramatically expanded the IC Group in 1987 and 1988. The initial offering of skid-steer loaders was followed by the introduction of several new lines, including the Dynaline telescoping-boom rough-terrain forklift, the MG-745 motor grader, and the GX-45 mini-excavator. IC Group sales increased eight-fold from 1986 to 1987, and the Company's plans called for IC Group sales to account for one-third of the Company's revenues by 1990.[238]

Gehl engineers introduced a variety of new skid loader attachments during the late 1980s that made the ubiquitous machines some of the most versatile on the market. In 1988, when the Indianapolis Motor Speedway repaved the venerable old two-and-a-half-mile racing oval in Speedway, Indiana, the job was entrusted to two Gehl SL4615 skid loaders with CP400 cold planers.[239]

The IC Group was created to market two product lines: skid-steer loaders and mini-excavators. Gehl Company's skid-steer loader line was expanding and the machines themselves were getting larger, as demonstrated by the 1985 introduction of the model SL6620.

Nearly twenty years after entering the compact construction equipment industry, Gehl expanded its line to include the mini-excavator. Five models, ranging from twelve horsepower to forty-six horsepower, were introduced in 1987.

In a setting that requires a safe surface for drivers, the Gehl skid-steer loader and cold planer were relied on in 1988 to resurface the Indianapolis Motor Speedway. Through its numerous sponsorships, Gehl Company remains active in auto racing and became affiliated with the Milwaukee Brewers baseball team in 2007.

John W. Gehl, vice president of corporate development and strategic programs, son of Mark Gehl, and grandson of the Company's founder, John Gehl, pointed out in early 1989 that "the IC Group continues to exceed market expectations. Our focus is on building a strong customer base through increased domestic distribution, new product offerings, and overseas expansion into specialized markets."[240]

Overseas Expansion

The overseas expansion into specialized markets was led by a licensing agreement in the People's Republic of China. Deng Xiaoping, the general secretary of the Chinese Communist Party, was presiding over a remarkable reform of the Chinese economy. Direct U.S. investment in China was rare at the time, but the Company's 1973 establishment of Gehl GmbH in Germany made Gehl a well-known name in Western Europe. Through Gehl GmbH, Gehl Company formed strong relationships with West German manufacturers and used those German contacts to open doors in the People's Republic of China.[241]

In the summer of 1987, John Gehl and Bernard Nielsen visited China and were amazed by the pace of development in the People's Republic of China. In 1988, Gehl played host to a trade delegation from China, and the Company signed a licensing agreement with the People's Republic to provide for the Chinese production of skid-steer loaders, the first such compact ever signed between a U.S. company and the People's Republic of China.

Gehl Company was the first skid-steer loader manufacturer to enter into licensing agreements with the People's Republic of China for the production of Chinese-built skid-steer loaders in China. John Gehl (center left) and his uncle, Al Gehl, are shown here with delegates from the People's Republic of China.

The move into China was ultimately unsuccessful, but it provided the groundwork for far greater success with international markets in the twenty-first century. Through the use of manufacturing, supply, and distribution licensing agreements, Gehl dramatically increased its export sales late in the decade. Sales were up 35 percent in 1989, and were two-and-a-half times the overseas sales the Company had reported in 1987.[242] During the 1990s, Gehl would capitalize on brand identity to grow its international presence.

Research and Development

As a consequence of its diversification into the light industrial and construction markets, Gehl began to invest heavily in research and product development. "We wanted to produce our products at a lower cost," Zadra said. "We also emphasized safer products because of the surge in products liability. Our insurance premiums were very high, at least $1 million a year. The Company was instrumental in setting up

Farmers rely on Gehl equipment to meet a variety of day-to-day needs, from harvesting crops and making feed to cleaning out barns and manure pits.

an industry captive insurance company to help us capitalize on our better-than-average safety record."[243]

The Company also experimented with doing other things in a different way. James ("Griff") Givens, vice president of manufacturing, reported in 1985 that the Company installed a new arc-welding robot to increase manufacturing efficiency and improve quality. The Company also introduced the use of high-solids paint, both to comply with government emissions regulations and to enhance the appearance of the Company's product lines.[244]

In 1985, Gehl initiated telemarketing, dealing directly with dealers from the West Bend offices, which eliminated the need for salesmen in remote areas, allowing salesmen to concentrate on their volume dealers. Kelly Moore was assigned to the telemarketing team in 1985. A native of Door County, Wisconsin, near the small town of Brussels, Moore earned his degree in marketing communications and journalism from the University of Wisconsin–Oshkosh in 1975 and went to work for an agricultural implement dealer in Door County the next year. Moore marketed Gehl products, and in 1978, he was asked to join Gehl Company as a service representative. Moore worked as a field service representative in LaCrosse from 1978 to 1980 and was a district sales representative in Waukesha from 1980 to 1984.

"In 1985, I started up the telemarketing program for the more remote areas of the United States," Moore said. "We had a special

A worker paints the beginnings of a skid-steer loader in the paint booth at the Madison, South Dakota plant.

Gehl sales were "flying high" by 1988, the year in which the Company welcomed shareholders to its annual meeting with this clever display of two of its IC Group products.

800-number, and we first started with the southeastern United States, North Carolina to Texas. We expanded in 1987."[245]

Going Public

The changes put into place after 1984 and 1985 positioned Gehl for some of the most spectacular growth the Company had ever enjoyed. Sales had reached nearly $100 million in 1979, the industry's record year, and then literally collapsed in the next three years, falling to $38 million in 1985. Net sales picked up in 1987, when revenues of $82.3 million were 53 percent ahead of the year before. In 1988, sales jumped another 51 percent, to $124.6 million, a 223-percent increase in three years.[246] The progress was even more pronounced in 1989, when Gehl reported net sales of $154.6 million, a 24-percent increase over 1988 and a third straight annual sales record.[247]

More important was the Company's growing record of profitability. In 1988, Gehl net income exceeded $4 million, an eye-opening turnaround from the $500,000 loss the Company had reported in 1985. In 1989, Gehl net income surged 23 percent to just over $5 million.[248] Total assets in

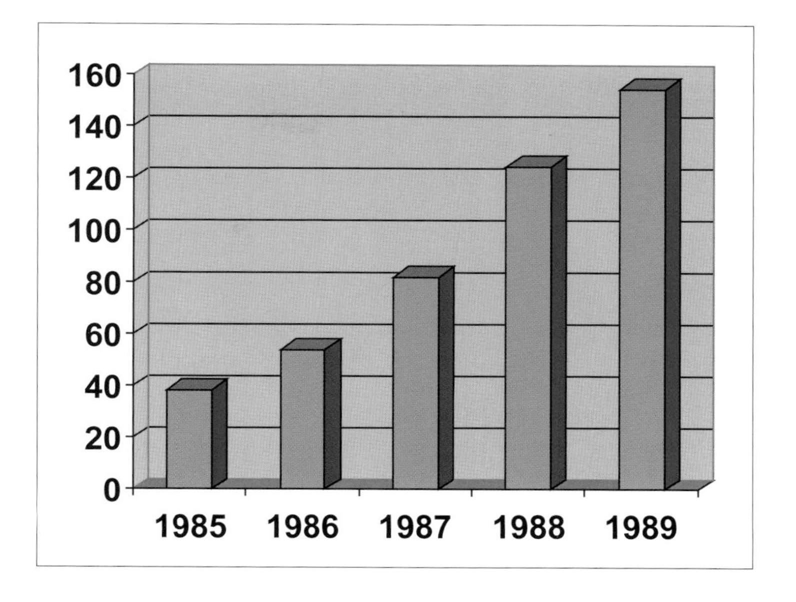

Net sales from 1985 to 1989 in millions, by year.

1989 were $150 million, and book value climbed to $12.27 a share. Capital expenditures for the year exceeded $3.1 million.[249]

Gehl capped one of the most remarkable five-year spans in the Company's long history when the firm's stock was listed on the NASDAQ exchange on November 21, 1989. The nearly two million shares of new common stock were oversubscribed, and Gehl used the $24 million raised in the offering to reduce borrowings under its revolving loan agreement.

The Berlin Wall had fallen just ten days before, and the free world was bright with promise. But the cyclicality that had plagued the agricultural equipment sector of the economy for almost a quarter-century was about to take a dramatic turn for the worse.

This announcement is under no circumstances to be construed as an offer to sell or as a solicitation of an offer to buy any of these securities. The offering is made only by the Prospectus.

November 21, 1989

2,200,000 Shares

GEHL®

GEHL COMPANY

Common Stock

Price $14 Per Share

Copies of the Prospectus may be obtained in any State in which this announcement is circulated from only such of the undersigned or other dealers or brokers as may lawfully offer these securities in such State.

Blunt Ellis & Loewi Incorporated

Robert W. Baird & Co. Incorporated

Bear, Stearns & Co. Inc.
The First Boston Corporation
Alex. Brown & Sons Incorporated
Dillon, Read & Co. Inc.
Donaldson, Lufkin & Jenrette Securities Corporation
Hambrecht & Quist Incorporated
Kidder, Peabody & Co. Incorporated
Lazard Frères & Co.
Merrill Lynch Capital Markets
Montgomery Securities
Morgan Stanley & Co. Incorporated
PaineWebber Incorporated
Prudential-Bache Capital Funding
Robertson, Stephens & Company
Salomon Brothers Inc
Shearson Lehman Hutton Inc.
Smith Barney, Harris Upham & Co. Incorporated
S.G. Warburg Securities
Dean Witter Reynolds Inc.
Wertheim Schroder & Co. Incorporated
William Blair & Company
Dain Bosworth Incorporated
A.G. Edwards & Sons, Inc.
Piper, Jaffray & Hopwood Incorporated
McDonald & Company Securities, Inc.
J.C. Bradford & Co.
The Chicago Corporation
First of Michigan Corporation
Stifel, Nicolaus & Company Incorporated
The Ohio Company
Prescott, Ball & Turben, Inc.
Frederick & Company, Inc.
B.C. Christopher Securities Co.
Cleary Gull Reiland McDevitt & Collopy Inc.
Rodman & Renshaw, Inc.
John G. Kinnard and Company Incorporated
Parker/Hunter Incorporated

Attributable to the initial public offering and the Company's 1989 earnings, stockholders' equity by the end of 1989 had increased to $71.4 million as compared to $42.9 million in 1988.

vari-sweep
980
GEHL

Chapter Eight

THE HARDEST *Years*

1990–1993

Hard times returned to Gehl Company and the nation's agricultural equipment manufacturers in the early 1990s. The last two years of President George H. W. Bush's single term in office were marked by the successful completion of Operation Desert Storm against Iraq's Saddam Hussein. But back home, the U.S. economy slipped into a stubborn recession that hiked unemployment rolls, kept interest rates high, and left manufacturers vulnerable to imports of everything from steel to farm equipment.

When Arkansas Governor Bill Clinton ran against President Bush in 1992, his rallying cry was simple. "It's the economy, stupid," Clinton told voters from Wisconsin to Wyoming.[253]

Like President Bush in the lightning-fast war against Iraq, Gehl Company enjoyed a very good year in 1990. The Company's first full year as a publicly held company brought record revenues of $175 million, up 13 percent from the $154 million reported in 1989. Net income topped out at $7.3 million, a 45-percent increase over the previous net income record of $5 million in 1989.[254]

Bernard Nielsen told shareholders that "in 1990 we turned an important corner. We completed a dynamic turnaround begun in 1985—and we entered our next growth phase aimed at becoming a strong global competitor."[255]

A Pause in the Up Cycle

Nielsen reported that 1990 began with "expectations that it would be one of the best for agriculture in a decade. A strong first half was fueled by healthy demand for our agricultural equipment both in North America and overseas."[256]

But what had started as a fantastic year for farmers and Gehl slowed perceptibly by the year's end. Milk prices declined precipitously in the fourth quarter, with a corresponding slowdown in farm equipment purchases. The IC Group, which had been growing at a strong double-digit annual percentage pace since its inception in 1984, eked out a 4-percent sales gain for the year following a softening of the construction business nationwide in mid-1990.[257]

Nielsen predicted caution in 1991, with farmers delaying the purchase of equipment. "We expect 1991 to be a pause in an up cycle, which was beginning in 1990, and project the industry to be strong in 1992 and beyond," he said, adding that the IC Group would be impacted by "continued recessionary conditions in 1991."[258]

The years 1991 and 1992 would prove to be the worst two years in the long history of Gehl Company. Sales would plummet during the two years, and the $19-million net loss

In 1990, Gehl Company made improvements to its line of skid-steer loaders and attachments, such as redesigning the model SL4615 skid-steer loader with a new engine for higher torque, better fuel economy, reduced noise, and fewer parts.

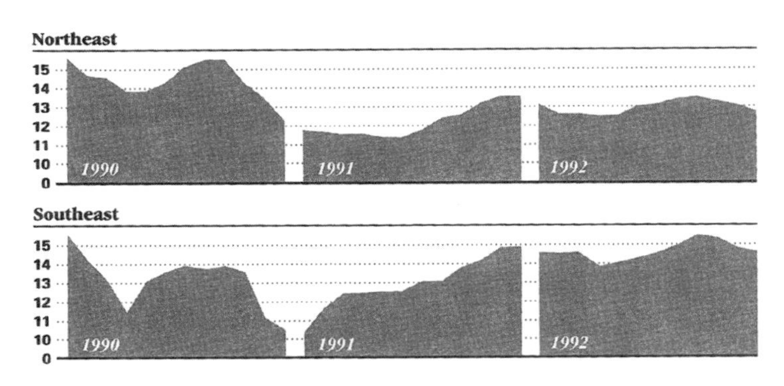

Milk prices per hundredweight. Starting in late 1990, the entire country experienced a drastic decline in milk prices, resulting in decreased market demand for agricultural equipment and a sharp drop in sales over the next several years for the Company.

in 1991 wiped out the Company's profits from the ambitious turnaround of 1985 to 1990. When the smoke finally began to clear in 1993 and 1994, the nation's agricultural equipment industry had irretrievably changed.

Managing Through Tough Times

The drop-off in milk prices in the fourth quarter of 1990 was symptomatic of wider changes taking place in American agriculture. The problems had begun to accelerate in the 1980s. Land values dropped throughout the decade, and tight credit made it difficult for the next generation of family farmers to purchase land and assets from their parents. Between the early 1980s and the mid-1990s, the number of farmers under the age of thirty-five declined dramatically in the U.S. Midwest. Nebraska, Iowa, and Minnesota lost more than 60 percent of young farmers during that period.[259]

Steps the Company has taken over the years to ensure quality customer service include: administering dealer satisfaction surveys, maintaining a dealer council, providing dedicated customer support representatives, and making a Web-based dealer portal available.

Meanwhile, the administration of President Ronald Reagan attempted to cut back on federal farm programs, arguing that many of the Roosevelt-era subsidies no longer made sense a half-century later.[260] The U.S. Department of Agriculture pointed out that some 70 percent of the 2.4 million U.S. farms produced less than $40,000 worth of commodities.[261]

The continuing strength of the U.S. dollar during the decade made American farm exports far less attractive, and commodity prices barely increased during the decade. As a result, many U.S. farmers found themselves barely able to afford the debt they had taken on to buy land and equipment during the 1970s. A growing number were forced to sell, often to corporate farm operations. Farms got bigger, and farming became more efficient and more productive.

The emergence of large, corporate farms led farmers to seek larger, self-propelled machinery instead of smaller, pull-type agricultural equipment.

A Loss of Market Share

The trend to corporate farming was particularly evident in livestock operations. Cattle feeding lots, hog-factory farms, and giant poultry operations increasingly began to replace the family farm as the 1980s ended and the 1990s began. And the reality for Gehl and other equipment manufacturers was that larger, more efficient corporate farms meant a loss of market share. With its reliance on haymaking, forage-harvesting, manure-handling, and feed-making equipment, Gehl was particularly vulnerable to the consolidation of livestock operations.

Joe Zadra recalled that "the industry went from smaller to larger farms, but Gehl did not have the self-propelled equipment to serve larger farms. In the 1950s, we were the first to introduce self-propelled forage harvesters. We sold them in Europe to custom operators who harvested crops for many small farms. They had some mechanical problems, so we dropped the product line. New Holland, Deere, and Fox all had self-propelled forage harvesting equipment in the 1980s. In the 1970s, there were 20,000 forage harvesters manufactured annually in the United States. We had 20 to 25 percent of the market, but in the late 1980s and early 1990s, the market just disappeared. We allied ourselves with dealers who didn't have a full line of implements."[262]

Joe Ecker retired as Gehl Company's vice president of marketing in 1992. He was witness to the beginning of a two-decade-long trend that would result in the virtual restructuring of the agricultural equipment market in the United States.

"A farmer needs one forage harvester, whether he farms fifty acres or 5,000 acres," Ecker said. "When farmers go out of business, they have an auction to get rid of their equipment. We had fewer and fewer and fewer farm customers."[263]

Gehl Company introduced the 552/553 Dynalift models in 1991. The easy maneuverability of the two new telescopic handlers made them highly suitable for space-restricted construction job sites.

A Perfect Storm

For Gehl, the changes taking place in the American farm economy collided with the Company's plans to spend money to meet its ambitious expansion targets. Those two factors came together in 1991 and 1992 with a resounding crash. A perfect storm of deteriorating business conditions and the accumulation of debt ravaged the Company's bottom line.

The Lebanon, Pennsylvania facility.

In hindsight, the 1990 decision to build a new manufacturing facility in Lebanon, Pennsylvania, was a mistake. Gehl broke ground for the new 170,000-square-foot plant in late 1990 and dedicated the facility exactly one year later.[264] The purpose of the new plant was to allow the Company's Agricultural Division to manufacture equipment closer to customer markets along the East Coast of the United States. Unfortunately, by 1991, many of those markets were drying up.

"We had done a study to determine where to build our new manufacturing facility," Mike Mulcahy said. "There was not enough capacity in West Bend or Madison. We built the new plant in Lebanon, Pennsylvania, which is just east of Harrisburg. It was a greenfield project built on land donated for a new community industrial park, and Gehl was the first to build there. We manufactured round balers and manure spreaders and assembled forage boxes."[265]

But the substantial capital investment in the new plant at Lebanon came at the worst possible time. "One of the problems we had was that we were making large investments in plant and equipment," explained Joe Zadra, who spent forty-two years with Gehl Company and retired from the Board as chairman in 1994. "We built a plant in Lebanon, Pennsylvania, because we thought that we could produce products closer to the Northeast market. We invested about $20 million in expanding production, but sales began to slide. The price of milk hit its lowest price ever in 1991. The economy was flat, which had a drastic impact over a short period of time in the early 1990s. The Company was caught with a very expensive infrastructure."[266]

Dave Ewald, a native of Madison, South Dakota, and a graduate of the nearby University of South Dakota at Vermillion, had gone to work at the Madison plant in 1984, when he was just out of college. The South Dakota economy was reeling at the time, and Ewald found out about the job posting at Madison when told about it by a classmate who had turned up her nose at working in rural South Dakota.

Ewald worked his way up through the Purchasing Department and became a software project manager at the Madison plant. In early 1992, he was named Yankton plant manager. "I always say that, had I known the financial condition Gehl was in at the time," Ewald said, "I never would have uprooted the family and moved to Yankton."[267]

Awash in Red Ink

The very expensive infrastructure associated with the Lebanon plant was almost immediately felt on the Gehl bottom line. In 1991, Gehl had $127 million in sales, a $47-

The Gehl Scavenger manure spreader was the primary product manufactured at the Lebanon, Pennsylvania plant.

Despite the Company's financial struggles at the time, Gehl engineers continued to work hard designing new equipment and improving existing models. In 1991, the 25 Series skid-steer loaders were introduced and offered many improvements over past models, including more operating capacity, increased dump height, stronger lift arms, and lower noise levels.

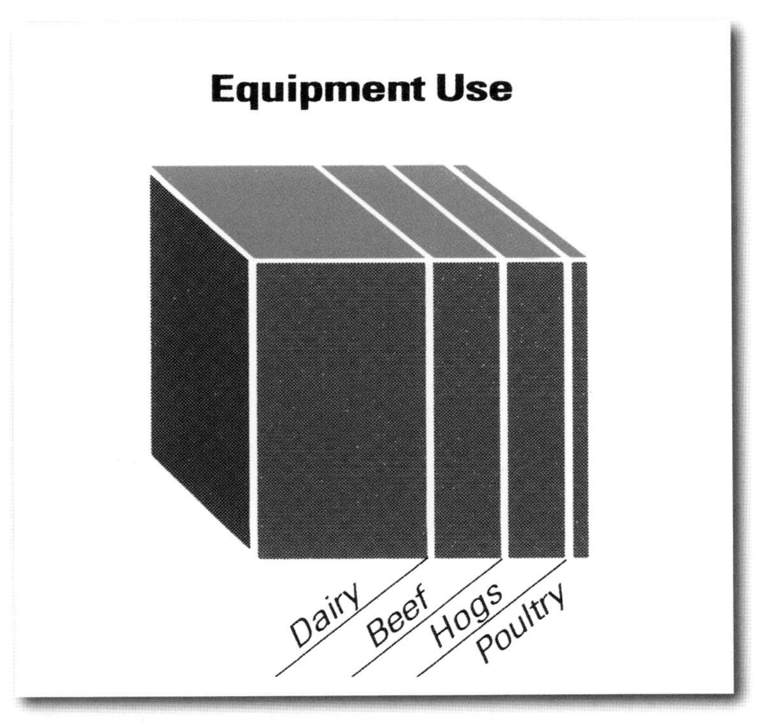

The customer base for Gehl agricultural equipment consisted primarily of livestock farmers.

million decrease from the year before. The lower revenues resulted in a loss of $19 million, which almost wiped out the $24 million that Gehl had received from its initial public offering just two years before. "We incurred a great deal of more expensive debt, with interest rates as high as 11 percent," Zadra said. "The debt restructuring became a real problem. We restructured $50 million with ITT. We paid down some of the senior debt because some of our banks were pressing us for payment."[268]

Gehl took what measures it could to stem the red ink. In late 1991 and early 1992, it shut down plants, reduced personnel, suspended the dividend, and cut operating costs. The Company eliminated management bonuses and cut compensation for all salaried employees.

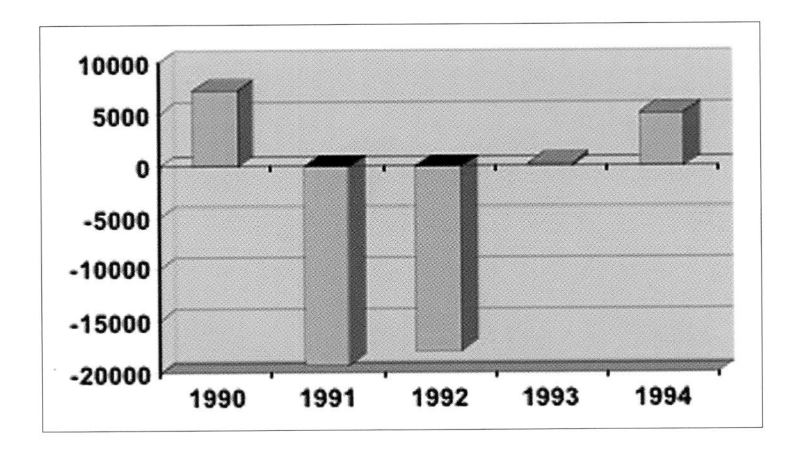

Net income from 1990 to 1994, in thousands, by year.

With milk prices at a thirteen-year low, sales of Gehl agricultural equipment continued to sag. Nearly 60 percent of the Agricultural Division's sales went to dairy farmers, which created inventory problems for Gehl and its dealers.

Bernard Nielsen remained optimistic that the Company could work its way out of trouble. He noted that Gehl had improved market share on most of its products in 1991 and told shareholders that the development or acquisition of twenty-six new products "will strengthen our market position and yield additional sales in 1992 and beyond."[269]

Default

But things didn't get better. By the third quarter of 1992, Gehl had lost more than $9 million during the first nine months of the year, on slightly improved sales. Nielsen explained in a shareholder release that "because industry-wide equipment inventory at dealers remain relatively high, we have had to meet the competition in terms of sales, discounts, and other incentives. We have also implemented specific programs to accelerate the reduction of older inventory at our dealers and increase cash flow over the balance of the year. The extent to which we have had to implement these special programs has been a critical unknown for us this year and has hurt our efforts to return to profitability."[270]

Increasing cash flow was particularly important in light of the Company's growing financial exposure. In October 1992, Gehl

Redlin, Browne & Company CPAs SC and
Wisconsin Manufacturers and Commerce proudly salute...

The 1990 Wisconsin Manufacturers of the Year

Wisconsin's manufacturing community is a world leader in innovation, skill and technology. With great pride, we announce the third annual Wisconsin Manufacturers of the Year

Winner	Category
Gehl Company West Bend	Large
ACCU-RATE, Inc. Whitewater	Medium
FOTODYNE, Inc. New Berlin	Small

Special Awards (for outstanding achievement in specific areas)

Special Award	Category
LEESON Electric Corp. Grafton	Applied Technology
A-C Equipment Services Corp. West Allis	Phoenix Award
Astronautics Corp. of America Milwaukee	Quality
Jagemann Stamping Company Manitowoc	Employee Development
SpecialtyChem Products Corp. Marinette	Environmental & Safery Management

Special Award	Category
Signicast Corporation Milwaukee	Customer Sensitivity
Paper Machinery Corporation Milwaukee	World-Wide Marketing
HM Graphics Inc. West Allis	Creative Commitment
Water Pollution Control Corp. Milwaukee	International Environmentalism
APPLETON Lamplighter Appleton	Employee Development

"Wisconsin manufacturing is continuing to exhibit strong growth. In the past three years, manufacturing employment has grown 11.8% in Wisconsin, three times faster than the national rate. In excess of 50,000 new manufacturing jobs were created here during that time. We're extremely proud to call some much-deserved attention to Wisconsin manufacturing, and to these superb representatives of Wisconsin's proud manufacturing tradition."

—*Michael J. Browne, President, Redlin, Browne & Company CPAs SC*

REDLIN BROWNE & COMPANY

WMC REPRESENTING WISCONSIN BUSINESS
WISCONSIN MANUFACTURERS & COMMERCE

The Business Journal

For nomination forms for the 1991 Manufacturers of the Year competition, write or call: Redlin, Browne & Company, 2775 S. Moorland Road, New Berlin, Wisconsin 53151, or call (414) 784-7900. Nominations are reviewed by an independent panel of judges.

Above: The 65 Series Gehl forage harvesters served the needs of the Company's row crop and feed operation customers. Designed for tractors up to 300 horsepower, the model CB1265 was the largest-capacity pull-type harvester when first introduced in 1993.

Left: In 1990, Gehl Company was honored by Wisconsin Manufacturers & Commerce (WMC) as Manufacturer of the Year in the Large Company category.

STOCK PRICES
1990 – 1993

	Price Range			
	1990	1991	1992	1993
First Quarter	$16.75 – 12.50	$9.75 – 5.88	$6.00 – 4.13	$3.88 – 3.00
Second Quarter	$16.75 – 14.50	$8.75 – 4.75	$5.25 – 3.38	$5.38 – 3.25
Third Quarter	$15.50 – 8.25	$6.75 – 5.25	$4.25 – 3.38	$6.88 – 4.25
Fourth Quarter	$ 9.75 – 6.50	$6.13 – 3.00	$3.50 – 2.25	$7.38 – 5.5
Year	$16.75 – 6.50	$9.75 – 3.00	$6.00 – 2.14	$7.38 – 3.00

Gehl Company stock prices from 1990 to 1993.

was in technical default with some of its lenders, and Nielsen and the finance team worked diligently to modify the Company's debt repayment schedules. An interim agreement signed that fall with lenders deferred principal payments of $3.5 million due at the end of the third quarter until the end of the fourth quarter.[271]

For Nielsen, the agreement to modify debt repayments was his last official act as Gehl president and CEO. On November 24, 1992, the Board of Directors announced that Nielsen had left the Company.[272]

For Joe Zadra, 1992 was a horrible year. "In 1992," he said, "we were in a very difficult position with the financial institutions. We had another loss of $18 million on sales of $152 million. Our stock price had dropped to between two dollars and three dollars per share."[273]

Complicating the Company's problems was the simple fact that as a publicly held company Gehl had to answer to Wall Street. When it was privately held, management could approach family shareholders and urge patience while problems were being worked out. But Wall Street, with its quarter-to-quarter mentality, demanded immediate solutions. And in mid-1992, there appeared to be few palatable immediate solutions on the horizon.

It was clear that Gehl had to undertake drastic measures to regain its footing. And the first drastic measure involved the appointment of an outsider, but a Gehl family member, as the new president and chief executive officer.

Fourth Generation

"Bill Gehl was on my Board during the whole workout period," Joe Zadra explained. "We had asked Bill to keep tabs on the place from the outside. Bill was with B. C. Ziegler and Company as executive vice president and chief operating officer. When Bernie left, we asked Bill to become president, and he told us that he would think about it. The next day, he agreed to take the job. It was one of my finer accomplishments. It was icing on the top of the cake. Gehl Company has been very successful and well-managed ever since."[274]

William D. Gehl's grandfather was Mike Gehl, one of the original "Gehl Brothers," and he grew up as the son of Daniel Gehl, a dentist in Milwaukee. "I graduated from Notre Dame," Bill Gehl said, "and it was the 'era' of Ara Parseghian. I went to the University of Wisconsin Law School and received my JD degree in 1971. But I had a yen for business. I went to the Wharton School of Finance and graduated with an MBA in finance in 1974. After I graduated from Wharton, I went to Miami to work with a corporate law firm. I also worked for Norwegian Caribbean Cruise Line as a purser on one of their cruise ships sailing from Miami. I did that for eighteen months and then returned to cold Wisconsin in 1975. I opened a law practice in West Bend as a sole practitioner in May of 1976."[275]

William D. Gehl joined Gehl Company as president and chief executive officer in 1992 and continues to serve as the chief executive officer and chairman of the Board.

Bill Gehl did a little bit of everything in his West Bend law office. He represented the local school board, handled divorces, and did title work for banks. He spent a year as a private attorney and then opportunity came knocking. "I got a call one day from Jack McCollow, the general counsel of B. C. Ziegler and Company," Bill Gehl recalled. "They were involved in investment banking and securities and wanted me to join them as assistant general counsel. I agreed to try it out for six months, and I stayed at Ziegler for fourteen years. I eventually became executive vice president and chief operating officer at B. C. Ziegler and Company."[276]

Joe Zadra asked Bill Gehl to join the Gehl Company Board of Directors in 1987. Bill Gehl helped establish corporate policy during the Company's late-1980s turnaround and was on the Board during the Company's precipitous decline in 1991 and 1992.

In November 1992, Bill Gehl joined the Company as president and chief executive officer. "The markets went down in 1991," Gehl said, "and at the same time, the Company was overextended. It found itself in a tenuous financial condition. The Board asked me to come in and help turn the Company around. I had a great deal of respect for the family legacy, but I didn't really have any aspirations to join the Company. In the back reaches of my mind, I always had a lot of respect for what Gehl Company stood for."[277]

Bill Gehl admitted to being torn. He enjoyed his work at B. C. Ziegler and could envision making his career there. "If circumstances hadn't been so dire," he explained years later, "I probably would not have come on board. Some people, including my wife, thought I was crazy. But I did view it as a once-in-a-lifetime opportunity to make a real difference. Time was of the essence."[278]

One of four U.S. patents awarded to Gehl product engineers in 1992 was for a net wrap attachment that increased the speed of baling operations and improved weatherability of the bale. All four patents were associated with hay tools, one of the largest markets for Gehl Company's agricultural implements.

Regaining the Bankers' Confidence

Bill Gehl recalled that the situation "was just exactly what I thought it was. There was financial stress, and the Company was operating under difficult conditions. Compromises needed to be negotiated with lenders. Loans needed to be restructured. Our first challenge was that the Company had defaulted on interest payments to senior lenders, and we had $110 million in debt outstanding. Our first task was to regain their confidence and get them to agree to a reasonable repayment schedule. We had underperformed, and we had to present an attainable business plan. The process started virtually immediately. In December 1992, we developed a business plan for 1993. Our goal was to pay a significant amount of the debt down and show a profit in 1993 of approximately $250,000."[279]

Bill Gehl had made it abundantly clear when he took the position that his strong suit was financial affairs. Accordingly, the Board agreed to hire a manufacturing specialist as executive vice president and chief operating officer. Victor A. Mancinelli came to Gehl Company from Cummins Engine Company in Columbus, Indiana, where he had held various senior management positions within the multi-site global manufacturer. A graduate of the University of Cincinnati, Mancinelli earned an MBA degree from Indiana University in Bloomington. He quickly took control of Gehl Company's manufacturing processes, imposing more rigorous cost controls and cost constraints.[280]

Bill Gehl was able to concentrate on rescuing the Company from its financial peril, secure in the knowledge that the Company's manufacturing operations were in capable hands. "Vic and I worked together closely for seven years from 1992 through 1999," Gehl said. "He was a tremendous help in turning things around. He was a very capable chief operating officer and a very bright guy who I enjoyed working with as the Company was restructured and returned to profitability. We

Gehl Company received the AE50 Award from the American Society of Agricultural Engineers for outstanding product innovation during 1992 for its new disc-mower conditioners.

made a great team."[281] With Mancinelli on board, Bill Gehl and Ken Kaplan, the Company's chief financial officer, were able to concentrate on restructuring the Company's debt. "The lenders gave us the necessary breathing room, and by September 1993, we had obtained new financing through ITT and repaid the senior lenders. We recovered, and we didn't look back. We even managed to eke out a small profit of $241,000 in my first year. It was the result of a good group of hardworking people who just needed to be redirected."[282]

Bill Gehl paid tribute to all those who worked hard with the turnaround team: "Bob Wright of Agricultural Sales and Jim Barnes of IC Sales were both instrumental in generating revenue and focusing the sales organizations on what needed to be done. We had an excellent team in South Dakota. Fran Janous headed operations at our Madison plant, and Dave Ewald was in charge of the Yankton plant. Fran was a very hands-on, take-charge guy, as is Dave. They were all key players."[283]

The Turnaround

By the beginning of 1993, Gehl Company was beginning its long recovery from the problems that it experienced in 1991 and 1992, a period during which the Company suffered combined losses of $37 million. The recovery was helped in no small part by improvements in the industries served by Gehl products, namely in construction and agriculture. The Company was also assisted in its recovery by growth in the international sector. Bill Gehl was happy to report in his letter to shareholders in early 1994 that the Company had posted a profit of $241,000. Though miniscule compared to some of the Company's best years in the late 1980s, it was a sign of real progress when compared to the $17.9-million loss in 1992.[284]

Agricultural sales were helped by lower interest rates combined with low farm debt. Those two factors contributed to the increased demand for agricultural equipment in 1993. Gehl Company's total agricultural equipment sales were $93,900,000 in 1993, a slight increase compared to $91,200,000 in 1992.[285] The jump in agricultural equipment sales demonstrated that Gehl had put itself in a position to benefit from the improving agricultural market through its cost reduction and inventory control programs, which were established in 1992. Gehl also improved its forage harvester line by introducing models that incorporated the latest technological improvements and by cutting supply-chain bottlenecks by quickly putting these new models into the hands of their dealers.

{ THE FLEXIBLE FACTORY, 1994 }

The mid-1990s ushered in an era of technological change that transformed manufacturing. Gehl was an early adopter of technology that made its employees more efficient and productive. Robotic welding, laser-cutting, and computerized inventory control became staples of the Company's manufacturing procedures during the 1990s, allowing Gehl workers on the factory floor vastly increased flexibility on how they produced agricultural and construction equipment.

The technological change is best seen at Gehl Company's plants in Madison and Yankton, South Dakota. Both were expanded dramatically after mid-decade to accommodate new technologies. Madison was expanded in 1994, 1998, 1999, 2000, and 2002. The size of the plant more than tripled from the early 1990s to 2002, from 80,000 square feet to 260,000 square feet.[289] Much of the expansion at Madison made room for the new technologies, particularly robotics.

"The size of the plant is the biggest change that I've seen," said Gary Rentz. "Technology is the other big change, and they're both related. All of our bigger items today use robotics. And laser cutting has been standard operating procedure for the last seventeen years, and maybe even longer."[290]

The story is similar at the Gehl plant in Yankton ninety miles south of Madison. Since 1997, Gehl has quadrupled the size of the Yankton plant, from just over 50,000 square feet to 216,000 square feet.[291] Much of the expanded space has been used for robotic welding stations and laser cutting operations.

The Yankton, South Dakota plant (top) and the Madison, South Dakota plant (bottom).

"Laser cutting is a huge deal," said Merrick Monaghan. "It literally changed the way you could design parts."[292]

At the Yankton plant, two large robots weld all the mainframes of the telescopic handlers that are assembled on the line. "Robotic welding enjoys a level of sophistication today that didn't exist years ago," said Dave Ewald, Yankton plant manager since 1992.[293]

While both South Dakota plants saw significant expansions and improvements, the West Bend plant, where the Company's agricultural products were manufactured, experienced change as well. It was upgraded in 2004 and 2005, including the installation of a new laser cutting system, to match the efficiency and production flow of the South Dakota plants.

Ewald noted that the Yankton plant continues to undergo technological upgrades that make its production lines more efficient and flexible. Gehl upgraded its paint system at Yankton in 2005, adding two new paint booths served by an automated monorail and a computerized five-stage washing system. Since 2000, the Company has made a significant investment in overhead lifting devices such as bridge cranes, so employees on the floor do not have to handle heavy parts. The plant also used a vendor-managed inventory system.[294]

The goal of the technology is to make it possible for Gehl employees to manufacture and assemble equipment in the least possible time at the lowest possible cost. Today, the technological changes have allowed Gehl and its employees an unprecedented flexibility in the way equipment is sent down the line.

"We used to build machines by lot, two hundred to three hundred at a time," explained Lyle Snider, formerly a technical customer service representative for Mustang in Owatonna, Minnesota. "Now, they build units one at a time. There are three or four different models going down the line at the same time. You put the order in, and it goes on down the line."[295]

For Yankton Plant Manager Dave Ewald, such flexibility has been a continuing source of amazement. "We build all of our products every day," he said. "There's very little batch manufacturing any more."[296]

The upgraded paint booth at the Yankton facility.

Efficiency is key, especially when a product is in demand. The skid-steer loader line in Madison kept pace with demand in the early 1990s by pumping out fully assembled machines in just thirty-three minutes.

Asphalt paver production has remained at the Yankton plant since the closing of the Georgia facility in 1994.

The construction industry began a slow turnaround as 1993 got underway, and the Company's compact construction equipment sales benefited as a result. "The year 1993 was one of transition and stabilization for Gehl Company," Bill Gehl said. "While the challenges were significant, I do not believe that the results adequately reflect the true potential of our business."[286] Retail demand for Gehl skid-steer loaders was strong, with sales increasing 18 percent over 1992. Another new piece of equipment that Gehl Company developed was the four-ton asphalt paver in an eight-foot-wide configuration. This new model was derived from the existing nine-foot PowerBox model and was a prime example of the Company's niche marketing strategy, making products that are durable and simple to operate while satisfying the customer's expectations.[287]

In 1991, the Company's new Gehl logo was introduced (above), replacing the logo that had been used since the mid-1970s (below).

The true potential of that business began to manifest itself during the mid-1990s. "The economy started to pick up in the mid-1990s," Bill Gehl explained, "and we were in a position to participate in the upturn. We did right-size the Company, although I hate to use that term. We reduced the overhead structure, and we have had some very good markets in the past fifteen years. Our goal has always been to not lose money when the markets turn down, and we've done that. That isn't always easy."[288]

Just how difficult it was to make money when the market turned became apparent late in the decade when the Company faced a hostile takeover attempt and later, in the twenty-first century, when it made the painful decision to abandon the agricultural equipment market that had been Gehl Company's core business since the earliest days of the Company.

6635 DXT
GEHL
Turbo
AirBOSS

Chapter Nine

RECOVERY AND *Rebirth*

1994–2008

Gehl Company clawed its way back to profitability in the mid-1990s and rode the wave of economic prosperity through the end of the century. In little more than a month's time, the Company survived a hostile takeover attempt and the temporary recession in the wake of al-Qaeda's attacks on the Twin Towers of the World Trade Center on September 11, 2001. Management made the momentous decision in April 2006 to exit the agricultural implement-manufacturing sector of its business and concentrate on manufacturing compact construction and industrial equipment. Agricultural implement sales had shrunk to less than 10 percent of the Company's overall revenue. As a result of the decision, Gehl closed its West Bend factory after 147 years of operation and focused on expanding manufacturing operations in South Dakota. Those facilities increasingly supplied European and foreign markets that placed a premium on the Gehl name. With the plummeting value of the U.S. dollar, by 2008, Gehl was sending nearly half of its production of skid loaders to dealers overseas.

Gehl Company's successes after the early 1990s and into the first decade of the twenty-first century were attributed to the leadership of Bill Gehl and the hard work of Gehl employees, some of whom were members of multi-generational families that had worked for the Company.

The Return to Profitability

By 1994, with the economic recovery taking hold across the nation and agricultural and construction markets going strong, the Company posted financial results that reflected the operational changes put in place in 1992 and 1993. Total sales increased 7 percent to $146.6 million compared to $137.2 million in 1993. Net income in 1994 rose to just over $5 million, a twenty-fold increase compared to $241,000 in 1993.[297]

Another positive aspect for Gehl Company in 1994 was its introduction of seven new product models for the agricultural market: four mixer-feeders, two round balers, and a forage box.[298] The launch of these products, along with other new machinery in both the agricultural implement and construction and industrial equipment markets, clearly demonstrated that product development would be the hallmark for steady improvement in all phases of the Company's operations during 1994 and in the years to come.

Gehl Company's recovery in 1993 and 1994 didn't escape the notice of Wall Street. By 1995, analysts and investors were commenting favorably on the Company's performance.

The popular Gehl mixer-feeder line included a large model with a proven four-auger system known for quickly blending and mixing feed ingredients into a consistent ration.

95 in '95

With Gehl Company solidly in the black by 1995, the Company was honored by the *Milwaukee Journal Sentinel*'s "95 in '95" listing of the top performers in terms of profit growth in the State of Wisconsin. Companies were ranked according to a two-year total return to shareholders through June 30, 1995. The article pointed out that Gehl ranked third in profit growth with a 286.6-percent increase from June 1994.

"Gehl stayed in the top dozen companies in the state for total return for the last two years at 63.41 percent," the *Journal* reported. "The year before, its two-year return was even higher, at 85.19 percent."[299] Earnings at Gehl hit an all-time high in the second quarter of 1995, jumping 112 percent to $3.6 million from $1.7 million in the same period of 1994. For the first six months of 1995, earnings were up 260 percent, to $5.4 million from $1.5 million a year earlier.[300]

The earnings figures clearly demonstrated that the strategy Gehl had outlined for its operations for 1992 and beyond had provided the basis for this dramatic recovery. But another factor that entered into the equation was that the demand for construction and agricultural equipment began to strengthen dramatically in 1995. In 1992, the market for skid-steer loaders hovered at around 27,000 units, but in 1995, the market increased to over 45,000 units in North America. In addition, the market for Gehl Company's rough-terrain telescopic forklifts became very strong in 1995 due to the increase in domestic construction spending.

Gehl Company's operation of the Yankton, South Dakota plant reflected the turnaround in sales during the early to mid-1990s. The Yankton plant started out manufacturing rough-terrain, telescoping-boom forklifts and motor graders. In 1994, Gehl closed its Lithonia, Georgia plant and transferred the manufacturing of asphalt pavers from Georgia to Yankton. The plant's workforce had shrunk to sixty people in 1992, and then doubled over the next three years as business picked up.

Merrick Monaghan, Yankton's manager of engineering, joined Gehl in 1988. He said even during the recovery, Gehl Company always had a light hand in the way it managed the Yankton plant. "Gehl always allowed a lot of autonomy," he said. "They didn't steamroll people. Gehl let the people in Yankton run it, and the telehandlers were always a pretty profitable line."[301]

Another improvement that contributed to the 1995 turnaround was enhanced inventory management. Kenneth Kaplan, Gehl Company's vice president and chief financial officer at the time, noted the Company's field inventory levels were "very competitive compared with other companies in the industry."[302] In addition, some of the plants were moving to a continuous-flow manufacturing method, which helped meet customers' Just-In-Time delivery requirements.[303] This manufacturing method would enable the Company to keep raw material and finished goods inventories reasonably low. Lower inventories also meant that more cash was available to be more profitably employed in such areas as debt reduction. That in turn would contribute to a corresponding decrease in interest expense.

As the second half of the 1990s began, Gehl found itself well positioned to take advantage of the dramatic upturn in

95 IN '95 MILWAUKEE JOURNAL SENTINEL	2 YEAR TOTAL RETURN ▼ 6/30/93 to 6/30/95	CLOSING STOCK PRICE 6/30/95	CLOSING STOCK PRICE 6/30/93	12 MONTH (6/30/95) REVENUES (000s)	12 MONTH (6/30/95) NET INCOME (000s)	12 MONTH (6/30/95) EARNINGS PER SHARE	MARKET VALUE (000,000s) 6/30/95
1. Lunar Corp.	140.43%	$28.25	$11.75	$ 44,572	$ 6,701	$ 1.14	$ 166.68
2. W.H. Brady Co.	96.28%	67.50	35.5	297,400	25400	3.50	492.75
3. David White Inc.	91.30%	11.00	5.75	16,552	(54)	(0.14)	5.54
4. Modine Manufacturing Co.	77.22%	36.75	21.50	943,800	69,600	2.27	1,124.55
5. Milwaukee Insurance Group Inc.	72.34%	20.25	11.75	122,819	4,555	1.10	84.04
6. Applied Power Inc.	71.92%	28.88	17.00	506,500	17,700	1.31	398.48
7. Harnischfeger Industries Inc.	71.47%	34.63	20.88	1,897,700	41,900	1.03	1,610.30
8. Manpower Inc.	68.37%	25.50	15.25	5,067,600	106,900	1.42	1,945.65
9. Regal-Beloit Corp.	66.41%	15.50	9.75	274,400	28,700	1.40	317.75
10. MGIC Investment Corp.	64.64%	46.88	28.75	549,943	183,053	3.08	2,777.59
11. Schultz Sav-O Stores Inc.	64.23%	22.50	14.00	441,300	5,700	2.27	56.25
12. Gehl Co.	63.41%	8.38	5.13	151,500	8,900	1.45	52.79

Gehl Company rounded out the top twelve companies recognized in the *Milwaukee Journal Sentinel*'s "95 in '95."

From jogging paths to driveways, golf cart paths to parking lots, Gehl PowerBox pavers are designed to handle all types of commercial and municipal applications.

In response to increased demand for skid-steer loaders in the mid- to late 1990s, the Company developed the 35 Series Gehl skid-steer loader line, represented in this photo by model SL6635, featuring durable run-flat tires. Prior to release, Gehl products are tested in the harshest environments, where Company engineers place the equivalent of a few thousand hours' worth of use on the machine.

Continuous-flow manufacturing is still very present in Gehl Company's production facilities as a result of the Company's constant commitment to improving manufacturing efficiency.

its markets. Realizing that they were in a cyclical business, the Company had positioned itself to make money in both up and down markets. "We never want to expose ourselves again to a down cycle," said Ken Kaplan.[304]

Adjusting the Product Line

With its 1993 debt refinancing firmly in place, Gehl turned to operational cost cuts that could help it increase profitability in the late 1990s.

Gehl started by focusing on its core products that were most profitable, eliminating low-profit models along the way. By 1995, the Company had cut 43 percent of its models, representing only 19 percent of its sales. The production cutbacks allowed the Company to put more of its cash into making products that generated higher earnings.[305] Gehl Company's personnel growth was also halted for a time. Employment was trimmed from nearly 1,400 workers to 900, positioning the Company to be at the size it needed to be profitable.

By the mid-1990s, three of the Company's four plants were what Gehl called "focused factories," which were dedicated to making just a few products instead of many. The Company also responded to a strong construction market, particularly a booming demand for skid loaders and rough-terrain forklifts, by emphasizing its models of this equipment.[306] By 1997, Gehl was expanding its two South Dakota factories to increase the manufacturing capacity of these products. Gehl expanded the Madison, South Dakota plant by one-third to meet the demand for skid loaders, and the plant at Yankton, South Dakota, was more than doubled in size to manufacture telescopic handlers and asphalt pavers.

The Company's oldest plant in West Bend, Wisconsin, also received attention during 1997, when Gehl implemented the final phase of continuous-flow manufacturing. This change resulted in increased productivity and improved efficiencies at the aging Wisconsin manufacturing facility.[307]

Gehl Company's balance sheet continued to receive attention through the Company's continued debt reduction program. Between 1991 and 1995, total debt was more than halved from $103 million to $47 million, which reduced interest expense by $4.4 million annually. As the stock price slowly rebounded by 1997 to five times its late-1992 value, Gehl Company could look ahead with confidence. The operational

By 1995, the models RB1475 and RB1875 were the only surviving Gehl round balers after drastic cuts were made to the Company's products.

Expansions at the Yankton plant have enhanced production by creating new assembly areas, improving the flow of work, and establishing more efficient shipping and receiving methods.

Building its first skid-steer loader in 1965, Mustang Manufacturing Company, Inc. is the second-oldest skid-steer loader manufacturer in the world. Today, the Mustang skid-steer loader line consists of nine models, including the model 2054 Mustang skid-steer loader shown here.

and financial changes that had been made were beginning to reshape the Company's future. The results were reflected in the Company's bottom line and healthy balance sheet. A solid foundation had been established to enable the Company to manufacture products of superior quality that were in high demand in the industry segments that the Company served.

Acquiring Mustang

Gehl Company's recovery had become so successful by 1997 that the Company could actually consider re-instituting the acquisition strategy previously abandoned following the 1992 recession. In 1997, Gehl purchased all issued and outstanding shares of Mustang Manufacturing Company, Inc. from its parent company, Brunel Holdings, PLC, thereby acquiring the Mustang line of skid loaders to complement Gehl Company's own line of skid-steer loaders.

"In 1997, we acquired what was left of Owatonna Manufacturing Company," explained Bill Gehl. "The strategy was to add a recognized line of skid-steer loaders. It gave us two brands of machines. Our goal was to broaden the Mustang-

Since acquiring Mustang in 1997, Gehl Company has significantly expanded the Mustang offering. The compact excavator group is the most extensive product line with eleven models available.

The original Mustang model 330 skid-steer loader.

produced line of skid loaders by introducing new products to Mustang dealers. Today, Mustang has a broad range of compact equipment, including skid loaders, compact excavators, telehandlers, compact track loaders, and all-wheel-steer loaders."[308]

Lyle Snider, a forty-year veteran of Mustang and Gehl in Owatonna, Minnesota, said, "The Mustang and Gehl brands were separate from 1997 on. At the time, we were running seventeen units a day through the Owatonna plant on a five-day shift. We had about 400 dealers, including a big dealer in Australia. Mustang shipped 300 to 400 units a year to them alone, and we also shipped to Europe and South America."[309]

Mustang primarily sold into the construction market, although skid-steer loaders are inherently versatile machines. Snider noted that Mustang "had a lot of units in the Iowa agricultural market. The Mustang 330 was the first hydrostatic drive skid loader with T-bar steering."[310]

Gehl Company's success, however, had not gone unnoticed in other quarters. Like many U.S. companies in the late 1990s and early twenty-first century, Gehl was about to experience the maelstrom of an attempt referred to by Wall Street and the financial media as a "hostile takeover."

A Corporate Game of Chicken

The turn of the millennium was an era in American capitalism that lionized the excesses of corporate raiders. The nation's financial media breathlessly followed the exploits of corporate raiders who launched hostile takeovers by assuring shareholders that they would get them more value. After the corporate raiders gained control of a company, they typically ejected the former management, issued bonds, sold off divisions, and pocketed huge profits, leaving the shareholders with only a shell of the company, which was often quickly acquired by a competitor.

Gehl Company's first experience with what some critics have called "vulture capitalism" came in 1997, about the time the Company was closing on its purchase of the Mustang operation. A Florida investor quietly amassed a block of nearly 250,000 common shares of Gehl common stock and 130,000 warrants.[311] The investor quickly upped his stake to 7.7 percent of the Company's shares in August 1997, paying as much as $20.38 a share.[312]

When Gehl discovered that the Florida investor was a former salesman

Market Place | Andrew Ross Sorkin

Caught in the sights of a takeover bid, Gehl turns to fight back.

IS the billionaire corporate raider Harold C. Simmons bluffing?

A construction equipment maker in West Bend, Wis., that is the target of a $100 million takeover campaign by Mr. Simmons is out to prove that his offer is simply a ploy to raise the company's stock price and his own fortune.

In a corporate game of chicken,

Business

Madison firm seeking success with blueberry beer 3D
www.jsonline.com/bym

WEDNESDAY, MAY 9, 2001 · MILWAUKEE JOURNAL SENTINEL

Texas group renews bid to buy Gehl Co.

Investors get billionaire to finance $18-a-share deal

By THOMAS CONTENT
of the Journal Sentinel staff

A Texas-based investment group has renewed its $18-per-share bid to buy the Gehl Co., and a billionaire backing the group said he has agreed to finance the deal.

The group did not raise the price of its offer for the construction and agricultural equipment manufacturer but left open that possibility, according to a filing Tuesday with the U.S. Securities and Exchange Commission.

When Gehl Co. rejected the group's original offer in December, the West Bend-based firm questioned whether the financing to back up the purchase was adequate.

In a letter made public in Tuesday's filing, billionaire Harold C. Simmons wrote, "I am prepared to finance directly or assist in arranging the financing necessary to complete the acquisition of the Gehl common stock."

William Gehl could not be reached for comment Tuesday. The company has said it does not believe a sale of the company is in the best interest of shareholders and would not provide them with long-term value.

The company has outlined a series of strategic initiatives that include new products, an expanded distribution network and the possibility of an acquisition of or alliance with an equipment-maker in Europe or North America.

In a letter to William Gehl on Friday, the suitors wrote, "We believe that this change indicates our commitment to promptly negotiate and conclude an acquisition of the company."

The suitors are two Dallas-based investment partnerships, Newcastle Partners, led by Mark Schwarz, and CIC Equity Partners, led by Paul DeRobbio. Together they control 346,000 shares of Gehl Co. stock, or about 6.5% of all the shares.

In seeking that the Gehl Co. board of directors engage in "serious discussions with us regarding the offer," they wrote, "we would be prepared to evaluate any information that the company has which would justify an adjustment in our offer."

The Texas group reiterated in the filing that a run-up in Gehl's

Please see GEHL, 6D

Business

METRO EDITION · TUESDAY, APRIL 24, 2001

Gehl's European partner buys 5.7% of company

The attempted takeover of Gehl Company was covered extensively by the media, from the *West Bend News* to a full-page spread in the Sunday *New York Times*.

The 80 Series round balers replaced the 75 Series round balers in 2001. The five new round balers all featured an exclusive variable open-throat power in-feed system, which enables customers to bale faster to beat the weather and build more bales while crops are at peak nutritional level.

While the executives were fighting off an attempted takeover in 2001, Company engineers were preparing to introduce a new line of skid-steer loaders. The 7000 Series took more than two years to design and develop and were labeled the biggest and strongest skid-steer loaders on the market.

for Michael Milken in the Beverly Hills office of Drexel Burnham Lambert, the high-flying junk bond firm of the 1980s, the Wisconsin company immediately began discussions with the investor, who had increased his ownership of Gehl shares to above 11 percent. In July 1999, Gehl re-purchased the investor's shares for $14.9 million, and in return the investor agreed not to invest in the Company for ten years.[313]

No sooner had the potential corporate raider from Florida been defused than Gehl was under attack from a Texas corporate raider whose exploits stretched back nearly twenty years. In 2000, billionaire Harold S. Simmons put together a nearly 5-percent stake in Gehl and, in August 2000, offered to buy the Company. During the 1980s, Simmons had engaged in a much-publicized battle with the management of Lockheed and had mounted a successful hostile takeover of Houston-based NL Industries.[314] Bill Gehl and the Board of Directors flatly refused to sell.[315] Just before Christmas, Simmons upped the ante and offered Gehl shareholders $18 a share for the Company.[316]

Gehl Company, like most of the rest of corporate America, had struggled with a weak stock market in 2000 in the wake of the collapse of the tech bubble on Wall Street. When Newcastle Partners, LP and CIC, LP, the Texas investor group affiliated with Simmons, made its late-2000 offer for Gehl, the Company's stock was trading at $15 per share, well below the $20-per-share Gehl stock fetched earlier in the year.

Gehl Company and Simmons' group sparred throughout the spring of 2001, with the battle dubbed a corporate "game of chicken" by the local media.[317] Bill Gehl even made the business pages of *The New York Times* that May. "The Company is not for sale," he said. "Period."[318]

Gehl Company postponed its April 2001 annual meeting until August to allow the Board of Directors time to chart a course of action, including possible acquisitions, strategic alliances, divestitures, a leveraged buyout, recapitalization, and the potential sale of the Company.[319]

The Company's stubborn refusal to knuckle under to Simmons and Newcastle Partners, LP seemed to be a futile effort when Gehl Company was forced to report sales and earnings below what Wall Street was anticipating.[320] In August 2001, Gehl again delayed its annual meeting, and the worst of the crisis appeared to be over the week after Labor Day, when Gehl stock closed at $18 on September 10, 2001.

The next day, al-Qaeda terrorists slammed jetliners into the World Trade Center in midtown Manhattan and the Pentagon

The Gehl compact excavator line was designed with user-friendliness in mind by offering precise performance and convenience features previously not found on excavators of similar size.

A cap displaying a "No Texas" symbol was worn by many employees during the attempted takeover in 2001.

in the most deadly attack on the U.S. mainland in history. When it reopened the next week, the stock market was down sharply. Gehl stock closed at $11.65 on September 26, 2001.[321]

Bill Gehl and the Company's Board were more than ever convinced that now was not the time to sell Gehl Company. In the months ahead, the Company weathered a potential proxy fight and shareholder suits. But in the end, the Company's uncompromising stand proved victorious. In early December, Simmons and his Texas group withdrew their hostile takeover bid after booking a sizable profit. They sold their stake in Gehl Company to Neuson AG, an Austrian manufacturing firm and a supplier to Gehl.[322]

Gehl Company and Neuson had begun partnering in 1999 when Gehl introduced a new line of mini-excavators made by the Austrian firm. Sold under both the Gehl and Mustang brands, the mini-excavators are rated from 1.5- to 12-metric-ton capacity and are outfitted with dozer blades and rubber tracks. All models featured an advanced high-output hydraulic system that allowed simultaneous multiple hydraulic functions without loss of power.[323]

Gehl Company's desire to stay independent and its commitment to that goal prevailed in the end. But the raid by Newcastle and CIC had exposed vulnerabilities in the Company's strategic plan. The Company would spend the early twenty-first century restructuring itself to compete in the new global economy.

Closings and Consolidation

At its September 27, 2001 Gehl Board of Directors meeting, the Board approved plans to streamline operations and cut costs. The Board voted to close the Lebanon, Pennsylvania plant and move production to South Dakota and West Bend. At the same time, Gehl transferred production of Mustang skid loaders from the underutilized Owatonna, Minnesota facility to Madison, South Dakota. The plant closings resulted in the elimination of one hundred jobs, or about 10 percent of the Company's workforce.[324]

Dan Miller, an Illinois native with manufacturing experience in South Dakota, joined Gehl as director of manufacturing operations in 2001, around the time the hostile takeover battle heated up. "When I first came to Gehl," he said, "they were beginning to move to a continuous-flow manufacturing method; however, inventory was still too high. If we could make money with this much inventory, what could we do if we reduced inventory and increased turns? We took $27 million out of inventory that first year and a half through improved processes in the supply chain and our manufacturing. Lean initiatives continued to drive out inventory as well as costs."[325]

Miller's first big project was closing the Lebanon, Pennsylvania and Owatonna, Minnesota plants. During the same period, Gehl expanded the Madison plant once and Yankton plant twice. Making compact equipment was a focus when Miller came in 2001. Sales of what the engineers called "red product" (agricultural equipment) were declining in 2001, and Miller tried to consolidate the red product into the West Bend East Plant to cut down on material handling and increase efficiencies.[326]

"We added a lot of technology in Yankton and Madison," Miller explained. "We added lasers and robotic welding. In manufacturing, you add technology and simplify the process. It reduces waste. We spend a lot of time analyzing processes to make our manufacturing leaner. We added a lot of metrics and measurement tools. If you can't measure it, you can't improve it. We put a lot of time into training and enhancing people's skills."[327]

The restructuring initiatives contributed significantly to the Company's improved results in 2004 and 2005. Gehl Company also looked increasingly to global partners for its continued growth in domestic and overseas markets. In 2004, the Company and the French company Manitou BF S.A., the world's largest manufacturer of telehandlers, entered into a strategic partnership. Gehl Company agreed to sell various models

Dan Miller, Gehl Company's vice president of manufacturing operations since 2005.

{ RIDING THE EXPORT MARKET, 2007–2008 }

As interest rates dropped in 2007 and 2008 to some of the lowest levels since World War II, the U.S. dollar took a dive in relation to many of the world's currencies. The dollar, which had been at parity with the euro in 2002, skidded to an exchange rate of $1.50 per euro.

For Americans, vacations in Europe became priced out of reach for all but the wealthiest. But for U.S. manufacturers, the weak dollar encouraged the sale of their products overseas. Oil shot up to $130 a barrel and beyond, and American manufacturers of farm and construction equipment suddenly found their products in greater demand in the rest of the world.

"We have been able to achieve significant growth globally," explained Malcolm Moore. "Approximately 51 percent of our skid-steer loaders will be shipped overseas in 2008. We've been trying to make sure that our employees understand that ours is truly a global market with an increasing portion of our sales as well as our supply chain coming from Europe and Asia. While we have an opportunity on the outbound, we also we need to be mindful that we have an exposure, in terms of the dollar [to] euro ratio, on the inbound."[340]

During the late 1960s, Gehl began marketing agricultural equipment to dealers in South America. Gehl Company's relationship with its European customers dates back thirty-five years. In 1973, Gehl Company set up

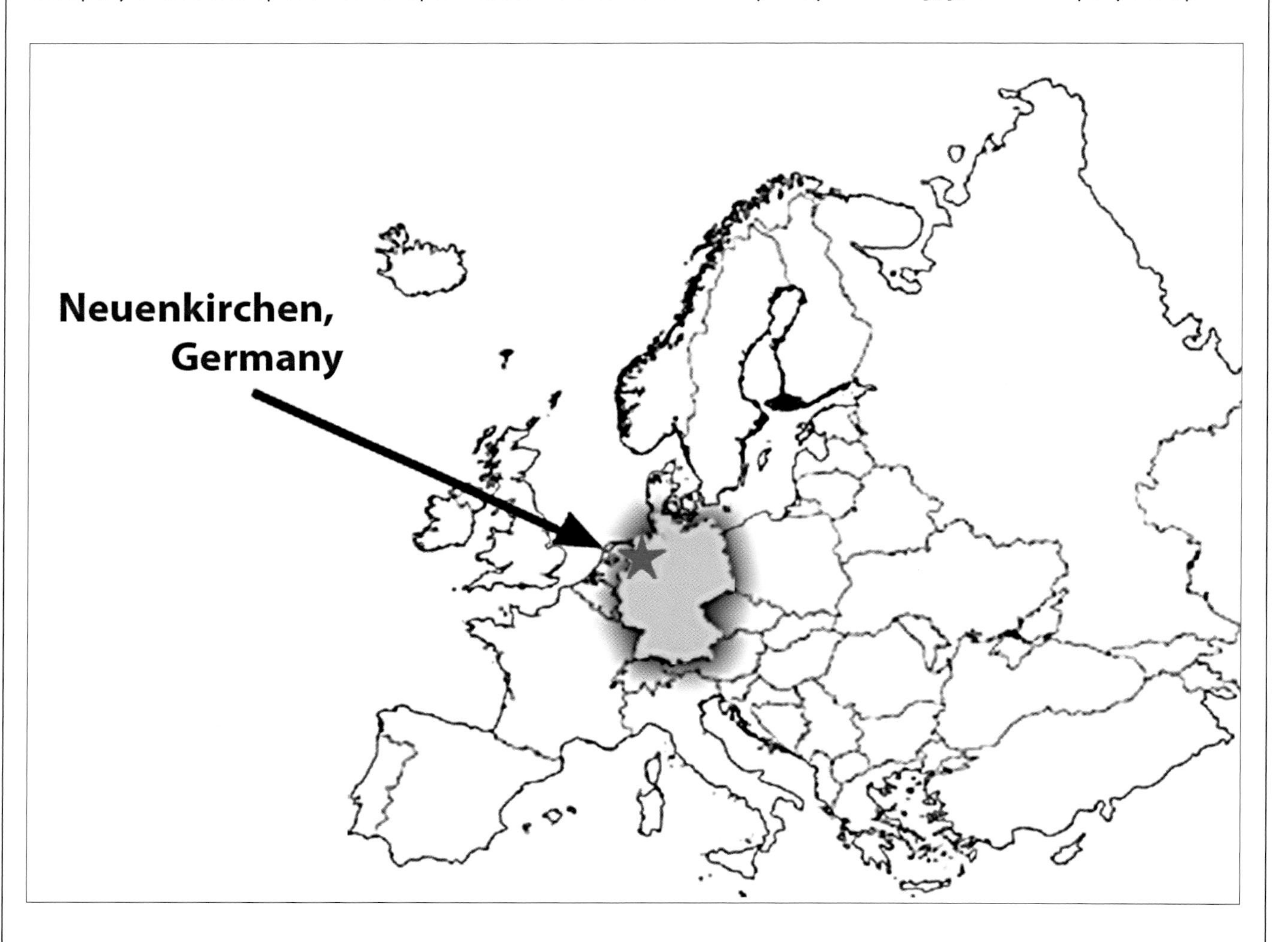

Gehl Company's 2002 purchase of Gehl Europe created new opportunities for growth. Located in Neuenkirchen, Germany, Gehl Europe handles the marketing, sales, distribution, and service of Gehl equipment throughout Europe, with the exception of a few key Mustang distributors covered directly by Gehl Company.

a company in Germany called Gehl GmbH to cover the European markets. Gehl hired a German national, Georg von Zech, who eased the transition and traveled to various countries for Gehl. Gehl Company owned 80 percent of Gehl GmbH's shares, and von Zech owned the remainder. Carl Gehl had been export manager, but he relinquished those duties to Joe Ecker because he no longer wanted to do the travel involved due to increasingly poor health.[341] In the 1980s, Gehl Company signed licensing agreements with the People's Republic of China. Although those agreements never met the Company's total expectations, they were a testament to Gehl Company's vision of global trade at a time when most U.S. manufacturers were hesitant to consider export licenses any farther away than Canada or Mexico. Later, in 1988, Gehl GmbH, Gehl Company, and a Chinese manufacturer started a joint venture producing skid-steer loaders in China.[342]

In 1985, Gehl Company sold its ownership in Gehl GmbH to a consortium led by von Zech. Eventually, by 2002, Gehl Company had repurchased 100 percent of the shares of Gehl GmbH, and then changed its name to Gehl Europe in 2003. The next year, Gehl Europe took over distribution of Mustang skid loaders in several European countries.[343]

In 2008, Bill Gehl noted that "today, having Gehl Europe under our roof gives us a great advantage to market in eastern and western Europe. Russia is vibrant with oil revenues, and we're selling quite a few skid-steer loaders in Russia. International trade is advantageous in combating the downturn in domestic housing construction. The agricultural market is booming. Our capability of developing business overseas is paying dividends today. It was a wise move years ago to be more global in our thinking. We are now reaping the rewards because more than half of our skid-steer loader product is being sold in international markets."[344]

Gehl Europe also markets certain equipment under the GEHLMAX brand name, which dates back several decades, as evidenced in this photo taken in a former Gehl Europe facility. One GEHLMAX product specific to Gehl Europe is the tracked dumper, a self-propelled carrying/dumping materials-handling machine unique to the European Union markets.

Just months before announcing the discontinuation of production of its agricultural implements, the Company introduced the new E-Series skid-steer loaders. This line of six new skid-steer loaders was designed with an optimized combination of performance, power, speed, and operator comfort.

Robotic welding stations enable engineers to program the welding units to create dozens of essential welds in precise locations.

of Manitou telehandlers through its dealer network in North America. Later in 2005, Gehl began manufacturing several of the Manitou models under license in its Yankton, South Dakota facility.[328]

Chairman and CEO Bill Gehl called the partnership with Manitou "an excellent opportunity for Gehl Company to broaden its product offerings to better serve the growing U.S. telescopic handler markets and to expand production at our Yankton, South Dakota manufacturing facility."[329] The relationship led to Manitou emerging as Gehl Company's largest shareholder.

The Manitou and Neuson partnerships and the growth of the compact equipment business in the early twenty-first century focused Gehl Company's attention on a key strategic question. Was the focus on construction equipment enough to sustain corporate growth and future development? How Gehl answered that question would define the Company's future direction.

Exiting Agricultural Implements

The news that Gehl had elected to exit the agricultural implement market, its core business for more than one hundred and forty years, came to West Bend on the first Monday of April 2006. Although a shock to the employees of the West Bend plant and to longtime Gehl equipment dealers, the news wasn't entirely unexpected.

Gehl had been wrestling with a declining dairy farming market for years. For the previous twenty years, there had been a continuous consolidation occurring in the U.S. dairy industry. Many of Gehl Company's small dairy farmer customers had either retired or sold their operations to larger producers. Consequently, the customer base for Gehl agricultural implement lines had declined dramatically.[330]

Much of Gehl Company's agricultural market had traditionally been in the dairy states of Wisconsin, Pennsylvania, and New York, but changes in the dairy farm industry had hit equipment manufacturers hard. As small dairy farms continued to undergo consolidation, the market for farm implements shrank accordingly. "In the 1985 U.S. Census of Agriculture, Washington County had 455 dairy farms," said Alan Linnebur, farm business educator for the University of Wisconsin Extension in Washington County. "In the 2002 U.S. Census of Agriculture, that number had dropped to 170."[331]

Between 1982 and 2002, the number of acres used for farming in Gehl Company's home of Washington County had decreased from 197,000 to 130,000, according to the U.S. Census of Agriculture. Farming acres had gradually shifted from dairy to cash crops. Because dairy farms grow forage, and participate in the type of agricultural activity that

Above: Expanding its product offering even further, in 2002 the Company began offering multiple all-wheel-steer loaders through the Gehl and Mustang brands with either conventional or telescopic booms and an industry-exclusive all-wheel-steer design.

Left: Gehl Company launched its CT-Series telescopic handler line in the spring of 2005 as a result of its strategic alliance with Manitou BF S.A. Gehl Company saw increased demand for telescopic handlers early in the 2000s, with 2005 experiencing a 67-percent increase in telescopic handler sales compared to the prior year.

Gehl Company has enjoyed a marketing partnership with IndyCar driver Graham Rahal since 2006. Gehl Company first became involved with racing in 1990, when it formed a partnership with the Toyota Grand Prix of Long Beach.

The model SL7810E Gehl skid-steer loader is the newest member of the E-Series family. Among its many features, the model SL7810E has the highest SAE-rated load capacity and highest lift height in the industry.

Gehl to end West Bend production

Despite 146 years of continuous service to the nation's dairy and livestock farmers, Gehl Company decided in 2006 to discontinue production of its agricultural implement lines.

was decreasing, the farm implement business suffered a severe decline. In Wisconsin, the number of forage crop acres dropped from 3.7 million in 1985 to 3 million by 2002.

Although the West Bend plant would be idled by the discontinuance of the farm implement business, Gehl planned to continue manufacturing at the two plants in South Dakota, focusing on compact equipment, including skid loaders, telescopic handlers, and pavers.[332]

The West Bend plant's 140 employees were allowed to keep their jobs for sixty days following the closing announcement. The State of Wisconsin's Department of Workforce Development formed a Rapid Response Team to begin interviewing employees before the announced closing date, to take profiles of each individual and assist with resumes, job searches, and job placement. The goal was to put these employees back into the workplace as quickly as possible.[333] Surprisingly strong demand for workers in the region helped those affected find new positions more quickly than anticipated.

Still, the closing was like a death in the family. Dale Luedtke, a Lomira, Wisconsin native who worked in the West Bend plant from 1972 until 2006, said he "took great pride in my job and pride in working for Gehl Company. I always had a lot of faith in the Company. A lot of good people were working here. It was a good place to work. I thought there'd be a future in agricultural products, but times change. It was hard for me to digest that there was no capacity to make money in this facility."[334]

While the area had recently experienced the closing of the Pressed Steel Tank Company in West Allis and Fisher Hamilton, LLC in Two Rivers, most observers felt that the Gehl closing would not have a long-term negative impact on the local economy. West Bend itself had already gone through the sale of the West Bend Company and the sale and closing of the Amity Leather manufacturing plant, but the state was prepared to do everything it could to lessen the effects of Gehl Company's layoffs and closing.

A Customer Is a Customer

Gehl Company was equally concerned with how its dealers and distributors would accept the change. But the Company had been preparing dealers for the primacy of construction equipment as early as the 1990s. Many of the dealers started to take on Gehl construction equipment at the time, and they were diversifying their customer base, which Gehl encouraged. Meanwhile, Gehl was diversifying its business, with an emphasis on providing compact equipment for both the agricultural and construction marketplaces. Exiting the agricultural implement business was a pivotal time for Gehl as a business, but the management team was certain that it was a strategy Gehl needed to follow.

Gehl Company's compact equipment product offering continued to strengthen with the Company's 2002 addition of compact track loaders. The new line was announced in conjunction with a strategic alliance between Gehl Company and Takeuchi Manufacturing Company, Ltd. to distribute compact track loaders in North and South America. The line is marketed under both the Gehl and Mustang brand names.

The majority of Gehl agricultural implement dealers had been selling Gehl skid-steer loaders for years, and some since the 1990s, which helped ease those dealers' transition into the compact equipment business.

Gehl Company continues to cater to the agricultural markets with its compact equipment offering.

Dan Keyes, Gehl Company's vice president of sales and marketing since 2000.

"Companies don't stick around this long by just being lucky," explained Dan Keyes, who came to the Company as vice president of sales and marketing in December 2000. "The agricultural implement markets were declining. Farms were changing, and we needed to provide products that were better suited to the changing needs of our dealers and their customers."[335] Keyes went on to say that the decision to exit the agricultural implement market "was not a decision that was made lightly or overnight. We recognized that it was time. It was a sound business strategy that was well received in the marketplace."[336]

As Gehl Company embarked on a new phase of its business, it became apparent that the way the Company supported current and future customers would be critical to its success. One of the last pieces of the plan was streamlining the Company organization by consolidating the sales and service support operations, explained Keyes. "We wanted to have a single point of contact for our dealers, to focus on taking care of all of their Gehl business regardless of the markets they served," Keyes said. "Initially, there was some consternation both internally and in the dealer network. However, when we communicated the strategy, the concern was addressed immediately. The dealers understood the changing marketplace and the decision we made that day was one that afforded both us and our dealers a more sustainable business partnership."[337]

The last piece Gehl put in place to solidify its position as a leader in the compact equipment market was the formation of a new subsidiary complementary to the Company's main product lines. "As we made the shift to the compact equipment business," explained Bill Gehl, "we created a separate company called CEAttachments Inc. It was launched to provide dealers with a complete line of attachments for compact equipment."[338]

Bill Gehl added that the painful decision to exit the agricultural implement market "changed the entire focus of the Company. It positioned Gehl as a compact equipment manufacturer for the agricultural and construction markets. We went from an agricultural implement manufacturer to selling more compact equipment to farmers and meeting material-handling needs on the farm. We've maintained that agricultural dealer distribution network and that gives us a great competitive advantage. It positions us for the next 150 years."[339]

CEAttachments Inc. markets high-quality attachments for compact equipment for use in a variety of applications, such as agriculture, construction, demolition, landscaping, nursery, mining, and foundries.

GEHL
GEHL

Chapter Ten

GEHL: THE NEXT *150 Years*

2008–2158

Gehl Company's twenty-first-century challenges—including a hostile takeover attempt, the economic fallout of the 9/11 attacks on the United States, and a restructuring that saw the Company exit the agricultural implement marketplace in 2006—were overcome through the vision of the Company's management team and the hard work of all the employees. And as an older generation of managers was retiring to make way for a new generation, Bill Gehl was able to assemble a new team, many from outside the Company, who brought new ways of doing business to Gehl.

The wave of new managers began in the spring of 1999, when Vic Mancinelli left Gehl to take the reins at CTB Corporation, an Indiana poultry equipment manufacturer. Mancinelli, who served as executive vice president of Gehl, had been an effective second-in-command to Bill Gehl, complementing the CEO's financial skills with his expertise in manufacturing.

Bill Gehl and the Board instituted a nationwide search to replace Mancinelli. The candidate who fit the bill was Chicago native Malcolm ("Mac") F. Moore. A veteran of twenty-five years of manufacturing and marketing experience, Moore had worked with such industry giants as FMC Corporation, General Signal, and the Merrill Lynch Corporation. Moore was living on the East Coast at the time, running a small investment group, and wanted to return to the Upper Midwest.

With his international equipment experience, particularly at FMC, Moore understood Gehl Company's need to restructure. "In the nineties, the majority of our sales were agriculturally based," he said, "including farm implements, but also skid loaders. People tended to see us primarily as a pull-behind (versus self-propelled) agricultural equipment manufacturer, which of course was our heritage. But the fact was we were increasing our strategic focus on products with higher growth potential for the compact construction equipment sector. Some of our traditional rural dealers were losing faith in the Company's ability to compete longer-term in the agricultural implement sector, and, in fact, our strength was gradually eroding because the American farm was undergoing consolidation. Typically, there were one-hundred-head dairy farms in 1999. Now there are some 4,000-head or larger corporate farms in California and Wisconsin. As farm size has increased, the trend is toward larger, self-propelled equipment, and that is the undisputed domain of much larger agricultural equipment suppliers, such as Deere, Claas, and New Holland."[345]

During the Company's one hundredth year of operation, nine pieces of the Gehl centennial line were painted a brilliant gold and put on display at the Pennsylvania Farm Show (above). Today, Gehl Company continues to increase its visibility in the compact equipment sector by taking part in the largest national and international tradeshows. Shown below is Gehl Company's 2008 display at CONEXPO in Las Vegas.

A New Management Team

Following the hostile takeover attempt in 2001, Bill Gehl assigned Moore the duty of spearheading the assembly of a new management team. "We are now more globally focused," Moore said, "and we have a new management team that is very growth-oriented. It took three years to build a new team, and then another couple of years to build some bench strength. We've been able to develop a better work force that is more externally focused. Today we compete successfully against Caterpillar, Terex, Komatsu, CNH Global, and the other big players. I am very proud of the management team's ability to have achieved significant profitable growth, both domestically and overseas, in company with some of the largest, most successful participants in the construction equipment sector."[346]

Two of Moore's immediate hires were Dan Keyes, who arrived as vice president of sales and marketing in late 2000, and Dan Miller, who came as director of manufacturing operations in 2001. Keyes, a Wisconsin native, worked for Case New Holland before joining Gehl. His work was cut out for him. "When I arrived at Gehl it was very apparent that there were two divisions with different ideologies within the Company," explained Keyes. "We had a long-standing and very successful agricultural division, along with an emerging construction division. The challenge we faced was streamlining the way we operated as a business while maintaining the levels of service to two separate customer bases."[347]

{ GEHL COMPANY'S 2008 OFFICERS }

William D. Gehl, *Chairman & CEO*

Malcolm F. Moore, *President & COO*

Ed Delaporte, *Vice President Information Technology*

Daniel M. Keyes, *Vice President Sales & Marketing*

Daniel L. Miller, *Vice President Manufacturing Operations*

James J. Monnat, *Vice President & Treasurer*

Michael J. Mulcahy, *Vice President, Secretary & General Counsel*

Brian L. Pearlman, *Vice President Human Resources*

Keyes' work to integrate the Company's sales and marketing function was critical in the eventual success of Gehl Company's 2006 decision to discontinue manufacturing agricultural implements and focus on providing other types of products to serve the changing agriculture landscape as well as the growing construction marketplace. Miller was responsible for ensuring that the Company could quickly and efficiently produce the increased number of construction equipment products to meet market demand and to replace the revenue from the abandoned agricultural equipment lines.

"We're the small fish swimming in the big pond, so we've got to be more flexible than the competition," said Miller, who was named vice president of manufacturing operations in 2006. "We have to be faster, more nimble, and more responsive to our customers and dealers. We have to be quick to the market with new products."[348]

As Chairman Bill Gehl took more responsibility for the overall strategic direction of the Company, he delegated financial management of the Company to new members of the Gehl team. Tom Rettler joined Gehl as chief financial officer in the summer of 2004 after working for Milwaukee-based WICOR, Inc. and Sta-Rite Industries for twenty years. During his four years at Gehl, the Wisconsin native helped the Company establish an asset-backed securitization facility, led a $46-million secondary stock offering, renegotiated Gehl Company's bank line of credit, and accessed the unrated commercial paper market to further reduce the cost of borrowings.

Rettler joined the Company just as its business was recovering from the downturn and recession caused by the September 11th terrorist attacks. With sales rising dramatically and the Company's balance sheet growing, Rettler quickly realized his principal challenge would be to secure adequate financing to support the Company's growth and market expansion plans. He would accomplish this by both reworking the Company's credit agreements and raising new equity capital from investors.

"From a growth standpoint," Rettler said, "we had the right product offering. The construction markets had seen a downturn, and the markets were recovering. We had new products. In 2003, the business reported $212 million in total revenue. By 2006, net sales had grown to $486 million. It put a lot of financial stress on the organization."[349]

Rettler and his team first arranged for additional borrowing capacity by entering into an asset securitization facility and then renegotiating the Company's corporate borrowing agreement. Using securitization, the Company sells to third-party investors installment sale notes that dealers use to finance the sale of equipment to their customers. With a new bank agreement, the Company significantly reduced its borrowing costs and gained flexibility in how to manage its business. "It took quite a bit of financial engineering," said Rettler, "but the Company was once again able to become an unsecured borrower. Our assets were no longer mortgaged under the debt agreement."[350]

Next came an additional sale of common stock to investors. This was Gehl Company's first offering of shares to investors since becoming a publicly traded company back in 1989. "In the fall of 2005, we did an equity offering," Rettler explained. "It was a primary equity capitalization. We did a full road-show to investors, and it was a learning experience. The markets started covering us. Before, Robert W. Baird was the only firm that covered us. We were able to pick up coverage from Sidoti and BMO and later Gabelli. We raised $46 million, which dramatically strengthened

The newest member of the line Gehl compact excavator family is the model CE283Z. Introduced in early 2008, the model CE283Z offers zero-tail-swing maneuverability, superior technology, and unparalleled agility.

our balance sheet. We went into the market at the closing price on the day of pricing. We didn't have to discount the offering."[351]

Rettler hired a number of new people into Gehl Company's Accounting and Finance departments. One of those Rettler brought on board shortly after starting at Gehl was Jim Monnat, a former colleague at WICOR. A Wisconsin native, and like Rettler, a longtime veteran of Sta-Rite and WICOR, Monnat brought to Gehl twenty years of corporate treasury and financing experience.

Other significant staff additions in short order included Andy Mohrfeld (tax director) and Shannon Van Dyke (now corporate controller) in 2005 and Michelle Gwin (director of internal audit) in 2006. "We had a very good group of people to work with," Rettler said. "What we needed to add were experienced staff who could take the Company up to the next level."[352]

Monnat came to Gehl after working for Wisconsin Energy Corporation. He immediately felt a sense of renewal when he began working at Gehl in January 2005. "For me, it was a culture shift, even though I have worked in a public company environment since 1982," said Monnat, who was named treasurer in 2005 and vice president and treasurer in 2007. "When I left Wisconsin Energy, I went from 6,000 employees to Gehl Company's 900 employees. I left a highly regulated industry to one driven entirely by market forces. Gehl is the smallest company I've ever worked for, but it is attractive because I was asked to be a member of the executive team. I think I fit in well with the management group, and I believe that my contributions are recognized and valued. Because Gehl is a public company, it is easy to see in our financial statements the changes we have made."[353]

Monnat assisted Rettler in restructuring the balance sheet and re-leveraging the business. The Company had previously tried to put in place a securitization financing facility. Within two months of his arriving at the Company, Monnat and his staff closed that transaction. Next came the revised bank credit agreement, the stock offering, and improvements to the securitization facility.

Monnat said that he has "very much enjoyed the last three years. I like the executive team, and the chain of command isn't too long. Nobody is more than three or four layers removed from the chairman-CEO. I like the interface between operations and sales, financing, and credit issues. It's a lot more than simply a corporate staff function."

The Dynamic Impact of Change

For Gehl Company, change has been a constant during the past century and a half. Perhaps the best illustration of the dynamic impact of change came as Gehl Company approached its 150th anniversary.

On September 8, 2008, Manitou BF S.A., a manufacturer and distributor of material handling equipment headquartered in France and the largest shareholder in Gehl Company since 2004, announced that it had signed a definitive agreement to acquire Gehl Company for thirty dollars per share. The $450-million purchase was realized through a tender offer for all outstanding shares of Gehl Company.[354]

Manitou, headquartered in Ancenis, France, is a world leader in rough-terrain material handling equipment, with revenues in 2007 of $1.8 billion. Like Gehl Company, Manitou serves three distinct markets: agricultural, industrial, and construction. Slightly more than half of Manitou's business involves the sale of masted forklifts, telescopic handlers, rotary telescopic handlers, truck-mounted forklifts, wheel loaders, and other equipment to customers in the construction industry.

The MC5 on a McCormick Farmall super-cub base; Marcel Braud (Junior) is at the controls.

In 2008, Gehl Company will open a new research and development center, constructed through the renovation of its former East Plant manufacturing facility in West Bend. The opening of a new corporate office facility, shown here to the right of the research and development center, will take place in 2009.

Manitou distributes its products through 600 exclusive and independent dealers in 120 countries, and its twenty-three global subsidiaries operate ten production facilities, including seven in Europe, two in China, and one in the United States. Manitou has approximately 2,800 employees worldwide.

Manitou traces its origin back to the early 1940s, when Marcel Braud (Senior), the grandfather of Marcel-Claude Braud, started the forerunner of today's business during World War II. Marcel Braud tragically was killed on August 5, 1944, the day the Americans arrived and liberated Ancenis from the Germans. His widow, Andrée Braud, formed the company that later became Manitou. Her son, Marcel Braud (Junior), joined the business in 1954, and his son, Marcel-Claude Braud, has served as president and CEO of the Manitou Group since 1998, making it three generations of Braud family involvement in Manitou.

Under the Manitou brand, the firm introduced and launched the first rough-terrain forklift truck in Europe in 1958, sold its 10,000th forklift truck in 1969, and signed a sales partnership agreement with Toyota in 1972 to distribute Toyota industrial forklift trucks in France.[355]

The combination of Manitou and Gehl Company provides obvious synergies in purchasing, research and development, manufacturing, and sales and marketing. The combined companies will establish a global leadership in the rough-terrain material handling equipment sector by building a presence for Manitou in the United States. And it will broaden the French parent company's product offering with the addition of skid-steer loaders, where Gehl Company holds a strong position among market leaders.

For Bill Gehl, the combination of the two companies is a culmination of the Gehl Company strategy to grow globally. "The combination of Gehl Company and Manitou," he said, "offers a substantial value to our shareholders today while affording our dealers and employees with future opportunities for continued success."

The More Things Change

The executive and management team that Bill Gehl and Mac Moore have established to lead the Company into another century of operation is heir to a proud tradition that stretches back to a frontier community on the west bend of the Milwaukee

The Company offers one of the broadest lines in the compact equipment industry as depicted in its 2008 catalog.

Gehl Company's corporate office building on Water Street in West Bend, Wisconsin, where the Company has been located for more than 115 years.

River. Much has changed in the past century and a half, but the challenges and opportunities that the people of Gehl face in the first decade of the twenty-first century are not much different than those faced by Gehl people in the 150 years since Louis Lucas built his foundry in the year that Abraham Lincoln and Stephen A. Douglas held their series of debates in neighboring Illinois.

The opportunities that Bill Gehl saw in focusing on compact equipment in 2006 were similar to the opportunities that his grandfather and great uncles saw one hundred years earlier when they borrowed money to rebuild their factory after the 1906 fire.

The challenges that Dave Ewald faces in starting a line for a new articulated loader at Gehl Company's Yankton, South Dakota plant are similar to those faced by Charles Silberzahn and his sons in designing and building a Hexelbank Silage Cutter at their West Bend factory in the nineteenth century.

The financial difficulties faced by Gehl in 1992 were a replay of the Company's difficult emergence from the ravages of the Great Depression in the late 1930s.

Dan Keyes' efforts to consolidate marketing initiatives and dealer networks in the early twenty-first century were reminiscent of Dick Gehl and, later, Joe Ecker's work to build the Gehl brand in the years after World War II.

Through it all, Gehl Company and its people have prospered by catering to the needs of its customer base and providing the marketplace with quality equipment at a reasonable price. Gehl has been successful because it has never forgotten its roots in the fertile native soil of Wisconsin.

In the spring of 2008, the Company introduced its model 340 articulated loader (below). The unit was designed in West Bend for the European market with a North American model to be introduced at a later time. The Gehl model RS5-19 telescopic handler (above) was designed in Yankton, South Dakota, and then brought to market in a similar fashion in early 2007.

Gehl Company's 2008 Board of Directors. From left to right: Richard J. Fotsch, Dr. Hermann Viets, William D. Gehl, Thomas J. Boldt, John T. Byrnes, John W. Splude, Bruce D. Hertzke, and Marcel-Claude Braud.

THE GEHL

Family Tree

Founders

John W. Gehl

Henry Gehl

Michael Gehl

Nick Gehl

Michael Gehl

Married Theresa Netzinger

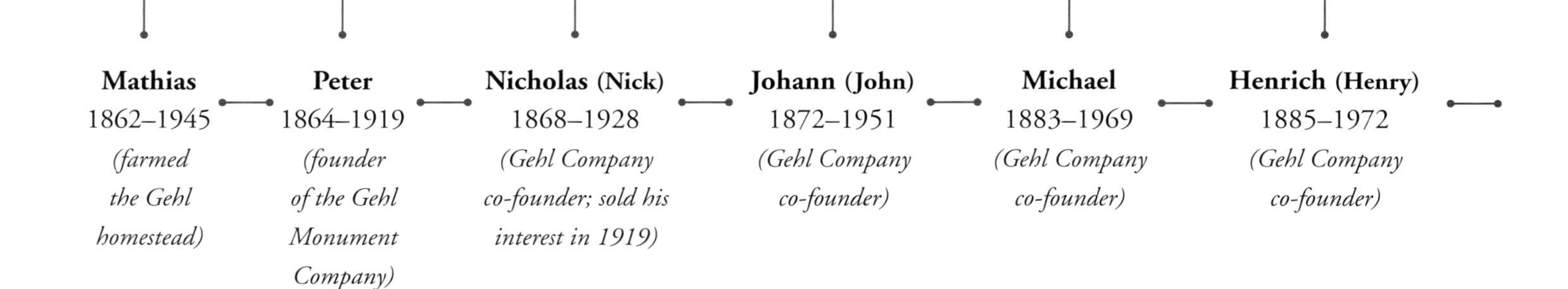

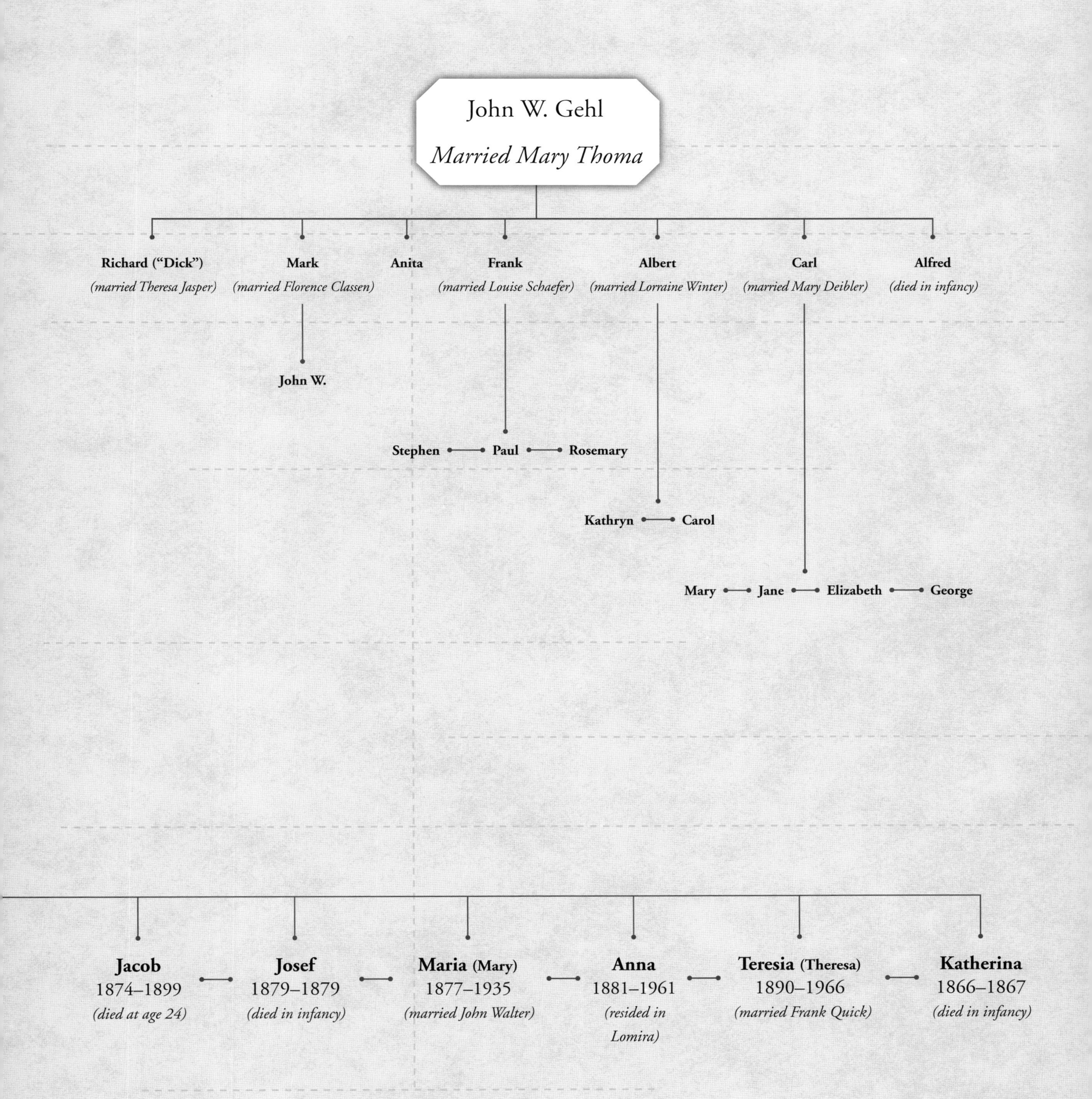
John W. Gehl
Married Mary Thoma
Richard ("Dick")
(married Theresa Jasper)
Mark
(married Florence Classen)
Anita
Frank
(married Louise Schaefer)
Albert
(married Lorraine Winter)
Carl
(married Mary Deibler)
Alfred
(died in infancy)
John W.
Stephen
Paul
Rosemary
Kathryn
Carol
Mary
Jane
Elizabeth
George
Jacob
1874–1899
(died at age 24)
Josef
1879–1879
(died in infancy)
Maria (Mary)
1877–1935
(married John Walter)
Anna
1881–1961
(resided in Lomira)
Teresia (Theresa)
1890–1966
(married Frank Quick)
Katherina
1866–1867
(died in infancy)

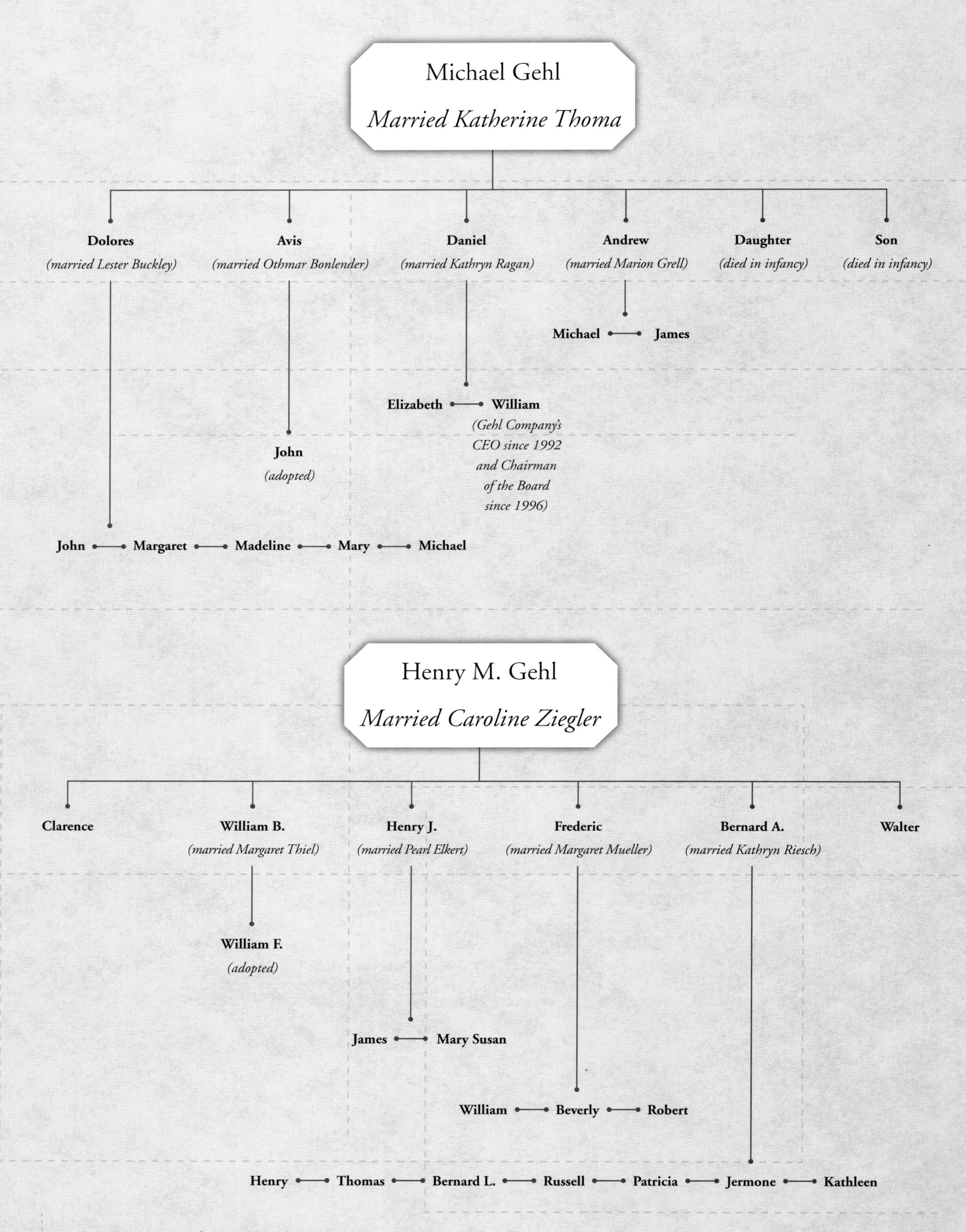
Michael Gehl
Married Katherine Thoma
Dolores
(married Lester Buckley)
Avis
(married Othmar Bonlender)
Daniel
(married Kathryn Ragan)
Andrew
(married Marion Grell)
Daughter
(died in infancy)
Son
(died in infancy)
Michael
James
Elizabeth
William
(Gehl Company's CEO since 1992 and Chairman of the Board since 1996)
John
(adopted)
John
Margaret
Madeline
Mary
Michael
Henry M. Gehl
Married Caroline Ziegler
Clarence
William B.
(married Margaret Thiel)
Henry J.
(married Pearl Elkert)
Frederic
(married Margaret Mueller)
Bernard A.
(married Kathryn Riesch)
Walter
William F.
(adopted)
James
Mary Susan
William
Beverly
Robert
Henry
Thomas
Bernard L.
Russell
Patricia
Jermone
Kathleen

Endnotes

1. Shiela Reaves. *Wisconsin: Pathways to Prosperity.* Northridge, California: Windsor Publications, 1988, p. 49
2. Ibid., p. 48
3. Case Corporation, www.en.wikipedia.org/wiki/Case_Corporation
4. Farm and Industrial Equipment Institute. *Men, Machines and Land.* Farm and Industrial Equipment Institute, 1974, p. 11
5. *Wisconsin: Pathways to Prosperity.* p. 57
6. Stewart H. Holbrook. *Iron Brew: A Century of Ore and Steel.* New York: The MacMillan Company, 1939, pp. 189–193
7. Western Historical Company. *History of Washington and Ozaukee Counties.* West Bend: 1881
8. Dorothy E. Williams. *The Spirit of West Bend.* Madison, Wisconsin: Straus Printing Company, 1980, p. 203
9. "West Bend," *Milwaukee Free Press,* January 23, 1913
10. Ibid.
11. "In West Bend Anno 1862," *West Bend Beobachter,* November 15, 1912
12. "The Freshet in Washington County," From the *West Bend Democrat,* quoted in the *Wisconsin State Journal,* June 9, 1858
13. Gehl Historical Files, Sketch No. 2, January 1948, p. 2
14. *Men, Machines and Land.* p. 11
15. Bill Beck and C. Patrick Labadie. *Pride of the Inland Seas.* Afton, Minnesota: Afton Historical Society Press, 2004, pp. 46–50
16. Ibid., p. 49
17. Gehl Historical Files, Sketch No. 1, March 1946, p. 1
18. Panic of 1873, www.en.wikipedia.org/wiki/Panic_of_1873
19. John R. Borchert. *America's Northern Heartland.* Minneapolis: University of Minnesota Press, 1987, p. 34
20. Wisconsin Historical Society. "Turning Points; The Rise of Dairy Farming," www.wisconsinhistory.org/
21. Norman K. Risjord. "From the Plow to the Cow: William D. Hoard and America's Dairyland," *Wisconsin Magazine of History,* Vol. 88, No. 3, Spring 2005, p. 44
22. William D. Hoard, www.en.wikipedia.org/wiki/William_D_Hoard
23. Company History: Kohler Company, www.answers.com
24. *The Spirit of West Bend.* pp. 204–205
25. "Silberzahn's fighting spirit aided in his success," *West Bend News,* April 18, 1985
26. Ibid.
27. "Gehl: 125 Years of Good Ideas," Unpublished Company Brochure, 1984, p. 1
28. J. F. Wojta. "Town of Two Creeks: From Forest to Dairy Farms," *Wisconsin Magazine of History,* Vol. 27, No. 4, June 1944, p. 435
29. Ibid., p. 428
30. Bill Beck. *50 Years of Cooperative Partnership: An Illustrated History of Hoosier Energy Rural Electric Cooperative.* Bloomington, Indiana: Hoosier Energy Rural Electric Cooperative, 1999, p. 3

31. Ibid., p. 5
32. O. A. Herbst, Peabody, Kansas. "No Repairs in Two Years." Herbst Testimonial from Silberzahn Catalog, February 4, 1913; Gehl Brothers Manufacturing Company, Annual Catalog of Silberzahn Ensilage and Feed Cutters, 1914, p. 25
33. A. M. Johnson, Doniphan, Nebraska. "Sells Other Cutter to Buy Silberzahn," February 17, 1913, Silberzahn Catalog, p. 25
34. C. F. Curtis, Ames, Iowa. "Prof. C. F. Curtis of the Iowa State College says," September 29, 1911, Silberzahn Catalog, p. 16
35. Gehl Family Genealogy in Gehl Genealogy File
36. Chronology of Luxembourg, www.rootsweb.ancestry.com/~luxwgw/luxchron.htm
37. Letter, from Georg von Zech, Locarno, Switzerland, to Carl Gehl, West Bend, Wisconsin, Gehl Genealogy File
38. *Wisconsin: Pathways to Prosperity.* pp. 52–53
39. Gehl Family Genealogy File
40. Mrs. Theresa Gehl, Unidentified Newspaper Clipping, Gehl Family Genealogy File
41. Gehl Family Genealogy File
42. Ibid.
43. Silberzahn Manufacturing Company, Annual Descriptive Catalog, 1900, p. 5; Gehl File, Washington County Historical Society
44. Silberzahn Catalog, p. 5
45. Ibid., p. 6
46. Ibid., pp. 12–32
47. John W. Gehl Profile, *History of Washington County.* West Bend, Wisconsin: 1911
48. Gehl Family Genealogy
49. "Charles Silberzahn: A maker of mighty farm machines," *West Bend News*, April 18, 1985
50. Ibid.
51. Annual Meeting of Shareholders, Berres-Gehl Manufacturing Company, May 11, 1904, Gehl Minute Books, p. 38
52. Special Meeting of Shareholders, Gehl Brothers Manufacturing Company, April 25, 1906, Gehl Minute Books, p. 44
53. "Gehl: 125 Years of Good Ideas," p. 2
54. Dale Luedtke Interview, p. 2
55. Annual Meeting of Shareholders, Gehl Brothers Manufacturing Company, June 13, 1907, Gehl Minute Books, p. 50
56. Gene Wendelborn Interview, p. 5
57. Peggy Lee Beedle. "Silos: an agricultural success story." University of Wisconsin Extension Service, n.d., pp. 1–13
58. Gehl Brothers Manufacturing Company, Annual Catalog of Silberzahn Ensilage and Feed Cutters, 1914, p. 20
59. Filling the Silo, op.cit., p. 44
60. Photo Caption, op.cit., p. 24
61. Farm Policy of the Twentieth Century, www.economics.about.com
62. Photo Caption, Gehl Brothers Manufacturing Company, Annual Catalog of Silberzahn Ensilage and Feed Cutters, 1914, p. 2
63. Gehl Brothers Manufacturing Company, Capital & Surplus at close of fiscal year, Gehl Minute Books, p. vii
64. Ibid.
65. History of Tractors, www.inventors.about.com
66. About Allis-Chalmers, Antique Tractor Shed, www.tractorshed.com
67. About John Deere, www.tractorshed.com
68. "Gehl Bros. Observes 100 Years of Progress," News Release, June 25, 1959, p. 5
69. Ibid.
70. Gene Smiley, Marquette University. The U.S. Economy in the 1920s, www.eh.net/encyclopedia/
71. U.S. History Encyclopedia. Dairy Industry, www.answers.com
72. "West Bend hosted Dairyman's Field Day 53 Years Ago," *West Bend News,* May 28, 1976
73. Gehl Brothers Manufacturing Company, Capital & Surplus at close of fiscal year, Gehl Minute Books, p. vii
74. Ibid.
75. Gehl Brothers Manufacturing Company, Capital Stock Authorization, February 23, 1937
76. Dividends Paid on Common Stock By Gehl Brothers Manufacturing Company, 1909 to 1922, Gehl Minute Books, 1922
77. *Three Generations: Gehl, 1859–1984*; "Gehl: 125 Years of Good Ideas," p. 10
78. Dwight Hoover. *A Good Day's Work: An Iowa Farm in the Great Depression.* Chicago, Illinois: Ivan R. Dee, 2007, p. 32
79. *Three Generations.* p. 10
80. Ibid.
81. Gehl Historical Files, Sketch No. 2, January 1948, p. 2
82. *Pride of the Inland Seas.* pp. 58–60
83. "Farm Implements," *Time* Magazine, April 5, 1926
84. F. W. Duffee. "Efficiently Filling the Silo," *Agricultural Engineering,* January 1925, pp. 4–12

85. Gene Wendelborn Interview, p. 2
86. Mayors, www.ci.west-bend.wi.us
87. Albert Gehl Interview, p. 1
88. Ibid., p. 3
89. Ibid.
90. John W. Gehl Genealogy, Gehl Family Genealogy File
91. "Death Takes R. M. Gehl," *Farm Machinery World*, January 1964
92. Albert Gehl Interview, p. 2
93. *Three Generations*. p. 12
94. *Wisconsin: Pathways to Prosperity*. p. 167
95. Ibid.
96. Lloyd and Don Theisen Interview, p. 1
97. "North Dakota Fun," *Time* Magazine, July 30, 1934
98. Lloyd Theisen Interview, p. 1
99. Ibid., p. 2
100. Lloyd and Don Theisen Interview, p. 2
101. Lloyd Theisen Interview, p. 2
102. Margaret Rosenthal. "West Bend, the city the depression missed," *West Bend News*, September 3, 1975
103. Gehl Brothers Manufacturing Company, Annual Meeting of Shareholders, December 12, 1933, Gehl Minute Books, p. 178
104. Ibid.
105. Ibid., p. 179
106. Gehl Brothers Manufacturing Company, Minutes of Special Stockholders Meeting, April 25, 1934, Gehl Minute Books, pp. 178–179
107. Ibid., p. 178
108. Hoover's Profile. The Ziegler Companies, Inc., www.answers.com
109. Phyllis Dausman. "They set the pace for industry," *West Bend News*, March 7, 1979
110. Ibid.
111. *Three Generations*. p. 11
112. Ibid., p. 8
113. *A Good Day's Work*. p. 60
114. *Three Generations*. p. 11
115. Ibid., p. 9
116. Ibid., p. 11
117. Terry Lefever Interview, p. 2
118. News Item, *Farm Industry News*, January 1975
119. "See New Three-Row Head For Gehl Choppers," *The Drovers Journal*, May 1974; "Gehl Offers Snapper Head Attachment for Choppers," *The Drover's Journal*, September 1974; and "New Knife-Sharpener For Forage Cutter," *New for Farm*, October 1975
120. Al Gehl Interview, p. 2
121. Ibid.
122. Digitally Recorded Oral History Interview with Gene Wendelborn, West Bend, Wisconsin, October 16, 2007, p. 1
123. Ibid.
124. Ibid., p. 2
125. Ibid., p. 3
126. Lloyd and Don Theisen Interview, p. 2
127. Ibid., pp. 2–3
128. Joe Ecker Interview
129. Al Gehl Interview, p. 3
130. Lloyd and Don Theisen Interview, p. 4
131. Ibid., p. 5
132. Ibid.
133. Ibid., p. 4
134. Bache & Company. Analyst Report, "Agricultural Equipment Industry," 1946, p. 1
135. "Gehls Breezing Along on Farm Tool Orders," *Milwaukee Journal*, June 13, 1948
136. Ibid.
137. Digitally Recorded Oral History Interview with Harold Gauger, West Bend, Wisconsin, October 16, 2007, p. 1
138. Ibid.

139. Gene Wendelborn Interview, p. 5
140. Digitally Recorded Oral History Interview with Laura Heisler, West Bend, Wisconsin, October 16, 2007, p. 1
141. 1908–1909, www.timelines.ws/20thcent/1908_1909.html
142. Albert Gehl Interview, p. 4
143. Ibid.
144. Ibid.
145. Ibid.
146. Ibid., p. 3
147. Gehl Company, News Release, August 2, 1967
148. "New Farm Machine Colors Announced by Gehl Company," *West Bend News*, November 22, 1966
149. Joe Ecker Interview, p. 2
150. Ibid., p. 1
151. Joe Zadra Interview, p. 1
152. Mervin C. Nelson. "Gehl Gives Farmer A Hand," *Milwaukee Sentinel*, March 4, 1968
153. Ibid.
154. Ibid.
155. Bruce L. Gardner. *American Agriculture in the Twentieth Century*. Cambridge: Harvard University Press, 2002, p. 6
156. Stephen J. Frese. "Comrade Krushchev and Farmer Garst: East–West Encounters Foster Agricultural Exchange," *The History Teacher*, Vol. 38, No. 1, November 2004
157. Joe Ecker Interview, p. 1
158. Ibid., p. 2
159. Ibid.
160. "Gehl Company History," Unpublished Company Brochure, 1972, p. 26
161. "Variable-Cut Transmission," *Feedstuffs*, February 29, 1964
162. "Gehl Company History," p. 27
163. Joe Ecker Interview, July 3, 2008
164. "What's New in Farm Machines For '67 Season?" *Drover's Journal*, February 3, 1967
165. News Item, *Prairie Farmer*, September 16, 1967
166. The Gehl Company, News Release, October 2, 1967
167. News Item, *Business Farming*, January 1969
168. News Item, *Southern Farm Equipment*, June 1969
169. News Item, *New For The Farm*, February 1969
170. News Item, *National Livestock Reporter*, September 1969
171. News Item, *Canadian Farm Equipment Dealer*, August 1970; Gehl Company, News Release, January 1, 1971
172. Marty Padgett. *Bobcat: Fifty Years of Opportunity, 1958–2008*. St. Paul, Minnesota: MBI Publishing, 2008, pp. 8–31
173. Joe Zadra Interview, p. 3
174. Joe Ecker Interview, p. 4
175. Hay Making and Handling Made Easier, October 2003, www.agrabilityproject.org
176. The History of Vermeer Corporation, www.vermeerag.com/history/
177. Joe Ecker Interview, p. 5
178. Ibid.
179. Joe Zadra Interview, p. 2
180. A Short History of the AIW, www.usw.org/
181. "Gehl Brother Dies," *Madison Daily Ledger*, October 23, 1974
182. "800 Workers Strike Gehl Co.," *West Bend News*, April 1, 1973
183. "Gehl Meeting Set; Strike 3 Weeks Old," op.cit., April 21, 1973
184. "Tentative Accord Reached in Gehl Bargaining," op.cit., June 7, 1973
185. "Gehl Union Okays Pact for 3 Years," op.cit., June 11, 1973
186. "Gehl Co. Announces Plant Expansion At South Dakota Site," op.cit., May 14, 1973
187. Ibid.
188. Federal Reserve Bank of San Francisco. Economic Research and Data, www.frbsf.org/publications/economics/letter/2004/el2004-35.html
189. Ibid.
190. Nebraska—History, www.city-data.com/states/Nebraska-history.html
191. Joe Zadra Interview, p. 4
192. Gary Rentz Interview, pp. 1–2

193. Ibid., p. 1
194. Joe Zadra Interview, p.4
195. Ibid.
196. Joe Ecker Interview, p. 6
197. Tape Recorded Oral History Interview with Laura Heisler, West Bend, Wisconsin, October 16, 2007, p. 1
198. Joe Zadra Interview, p. 2
199. Richard J. Semler. "Data Systems & Administration," Gehl 1985 Annual Report, p. 10
200. Joe Zadra Interview, p. 3
201. Ibid.
202. "Products Boast Refinements, Innovation," Gehl 1981 Annual Report, p. 7
203. "Gehl Introduces New Generation Round Balers," *The Gehl Leader*, January 1981, p. 4
204. Ibid.
205. "New Products are State-of-the-Art," Gehl 1984 Annual Report, p. 6
206. "Disc Mower Conditioners," op.cit., p. 7
207. "To Our Shareholders," op.cit., p. 3
208. "Tom Lyons, Gehl to close 6 more weeks this year," *West Bend News*, September 27, 1982
209. "Gehl production lines rolling again," *The Milwaukee Journal*, May 6, 1982, p. 7C
210. Ibid.
211. "Kamps resigns from Gehl Company," *West Bend News*, December 15, 1983
212. "Kamps named new Kasten president," op.cit., October 16, 1984
213. Terry Lefever Interview, p. 4
214. Ibid.
215. Ibid.
216. Ibid.
217. "Finance," Gehl 1985 Annual Report, p. 10
218. Ibid., p. 4
219. "Sales of Finance Contracts Receivable," Gehl 1989 Annual Report, p. 18
220. Mike Mulcahy Interview, p. 1
221. Joe Zadra Interview, p. 5
222. "A Strategic Perspective, 1985–1989," Gehl 1989 Annual Report, p. 4
223. Ibid., p. 5
224. "New Gehl president named," *West Bend News*, April 30, 1985
225. Mulcahy Interview, p. 3
226. Joe Zadra Interview, p. 5
227. Merrick Monaghan Interview, p. 1
228. "Manufacturing," Gehl 1985 Annual Report, p. 8
229. Mike Mulcahy Interview, p. 3
230. Joe Zadra Interview, p. 5
231. Ibid., p. 2
232. Lyle Snider Interview, p. 3
233. Ibid., p. 4
234. Mike Mulcahy Interview, p. 4
235. "To Our Shareholders," Gehl 1985 Annual Report, p. 3
236. Ibid.
237. Letter from the President, Gehl 1987 Annual Report, p. 4
238. "IC Group," Gehl 1987 Annual Report, p. 9
239. "Expanding beyond market growth," Gehl 1988 Annual Report, p. 14
240. Ibid., p. 13
241. Joe Schoenmann. "Gehl Co. welcomes Chinese partnership," *West Bend Daily News*, n.d.
242. "To Our Stockholders," Gehl 1989 Annual Report, p. 3
243. Joe Zadra Interview, p. 5
244. Manufacturing, Gehl 1985 Annual Report, p. 8
245. Kelly Moore Interview, p. 1
246. "Financial Highlights," Gehl 1988 Annual Report, p. 3
247. "Financial Highlights," Gehl 1989 Annual Report, p. 1

248. Ibid.
249. "A Strategic Perspective, 1985–1989," Gehl 1989 Annual Report, p. 5
250. "To Our Shareholders," Gehl 1989 Annual Report, p. 2
251. "Operations Review: Agricultural Division," op.cit., p. 9
252. "Operations Review: IC Division," op.cit., p. 11
253. Chris Hegedus and D. A. Pennebaker, directors. *The War Room.* Cyclone Films: 1994
254. "Financial Highlights," Gehl 1990 Annual Report, p. 1
255. "To Our Shareholders," Ibid., p. 2
256. Ibid.
257. Ibid.
258. Ibid., p. 3
259. Initiative 300 and Beginning Farmers, www.I300.org/factsheets.htm
260. George J. Church. "Real Trouble on the Farm," *Time* Magazine, February 18, 1985
261. Ibid.
262. Joe Zadra Interview, p. 4
263. Joe Ecker Interview, p. 7
264. "Gehl Company Opens New Manufacturing Facility, Celebrates the Future of Agriculture," News Release, November 25, 1991
265. Mike Mulcahy Interview, p. 4
266. Joe Zadra Interview, p. 6
267. Dave Ewald Interview, p. 1
268. Joe Zadra Interview, p. 6
269. "Letter to Shareholders," Gehl 1991 Annual Report, p. 2
270. "Weak Markets Slow Gehl Company Recovery," News Release, October 30, 1992
271. Ibid.
272. News Release, November 24, 1992
273. Joe Zadra Interview, p. 6
274. Ibid.
275. Bill Gehl Interview, p. 1
276. Ibid.
277. Ibid., p. 2
278. Ibid.
279. Ibid.
280. "Victor A. Mancinelli Named CEO of CTB International Corp.; Chris Chocola to Serve as Chairman of the Board," *Business Wire*, March 10, 1999
281. Bill Gehl Interview, p. 2
282. Ibid.
283. Ibid., p. 3
284. "Letter To Shareholders," Gehl 1993 Annual Report, p. 2
285. Ibid.
286. Ibid., p. 3
287. Ibid., p. 4
288. Ibid., p. 3
289. Gary Rentz Interview, p. 2
290. Ibid.
291. Dave Ewald Interview, p. 3
292. Merrick Monaghan Interview, p. 3
293. Dave Ewald Interview, p. 4
294. Ibid., p. 5
295. Lyle Snider Interview, p. 5
296. Dave Ewald Interview, p. 5
297. "Letter To Shareholders," Gehl 1994 Annual Report, p. 2
298. Ibid, p. 5
299. "Gehl Co. plows new path on way to profitability," *Milwaukee Journal Sentinel*, October 9, 1995
300. Ibid.
301. Merrick Monaghan Interview, p. 2

302. "Gehl Co. plows new path on way to profitability."
303. Ibid.
304. Ibid.
305. Ibid.
306. "Gehl-force makeover," *Milwaukee Journal Sentinel*, June 9, 1997
307. "Chairman's Message," Gehl 1997 Annual Report, p. 3
308. Bill Gehl Interview, p. 3
309. Lyle Davis Interview, p. 4
310. Ibid.
311. "Investor buys stake in Gehl," *Bloomberg News*, June 10, 1997
312. "Gehl Co. Stake Raised by Investor James Dahl to 7.7% from 5.9%," *Bloomberg News*, August 14, 1997
313. "Gehl buys out dissident investor," *Milwaukee Journal Sentinel*, July 10, 1999
314. "Caught in the sights of a takeover bid, Gehl turns to fight back," Market Place, *New York Times*, May 9, 2001
315. "Gehl rebuffs offer," op.cit., September 9, 2000
316. "Group offers 67% premium for Gehl," op.cit., December 23, 2000
317. "Game of chicken," *West Bend Daily News*, May 9, 2001
318. "Caught in the sights of a takeover bid, Gehl turns to fight back," Market Place, *New York Times*, May 9, 2001
319. "Gehl board of directors to chart a course," *West Bend Daily News*, May 9, 2001
320. "Firms express interest in Gehl," *Milwaukee Journal-Sentinel*, June 27, 2001
321. "Gehl decides against sale of company," op.cit., September 27, 2001
322. "Pursuer of Gehl sells its stake," op.cit., December 8, 2001
323. "Gehl Introduces New Line of Excavators," News Release, May 24, 1999
324. "Board Rejects Sale Proposal, Plans Consolidation, Layoffs," *The Wall Street Journal*, September 28, 2001
325. Dan Miller Interview, p. 1
326. Ibid.
327. Ibid., p. 2
328. "Gehl Company," Gehl 2004 Annual Report, p. 6
329. "Gehl Company and Manitou Announce Strategic Alliance In the U.S. Telescopic Handler Markets," News Release, July 22, 2004
330. "Dealt a Bad Blow," *The West Bend Daily News*, April 4, 2006
331. Ibid.
332. Ibid.
333. "Gehl workers to receive employment help," *The West Bend Daily News*, April 4, 2006
334. Dale Luedtke Interview, p. 2
335. Dan Keyes Interview, p. 2
336. Ibid., p. 3
337. Ibid., p. 2
338. Bill Gehl Interview, p. 2
339. Ibid., p. 1
340. Malcolm Moore Interview, p. 2
341. Joe Ecker Interview, p. 3
342. "Short History of GmbH/Gehl Europe GmbH," Unpublished Company Brochure, 2008
343. Ibid.
344. Bill Gehl Interview, p. 3
345. Malcolm Moore Interview, p. 2
346. Ibid., p. 3
347. Dan Keyes Interview, p. 2
348. Dan Miller Interview, p. 2
349. Tom Rettler Interview, p. 2
350. Ibid., p. 3
351. Ibid., p. 4
352. Ibid.
353. Jim Monnat Interview, p. 1
354. "Gehl Company Announces Agreement to be Acquired by Manitou BF S.A. for $30 per Share," News Release, September 8, 2008, p. 1
355. History, www.gp.manitou.com

ABOUT THE

Author

Writer and historian Bill Beck has more than two decades of experience writing about business history. He wrote his first history for Minnesota Power in 1985 and has more than seventy-five published books to his credit. Recent books include illustrated anniversary histories for Telsmith, Inc. in Mequon, Wisconsin, and Arkema, Inc. in Carrollton, Kentucky. He is currently writing a 150th anniversary history for Gardner-Denver Corporation in Quincy, Illinois.

Beck is a 1971 graduate of Marian College, Indianapolis, and did graduate work in American history at the University of North Dakota. Beck started Lakeside Writers' Group twenty years ago, following ten years as a reporter for newspapers in Minnesota and North Carolina and seven years as the senior writer in the public affairs department at Minnesota Power & Light Company in Duluth, Minnesota.

1859

Gehl Family

Top Row: (left to right) — *John, Peter, Michael (father of the company founders), Theresa (his wife), Mathias, Nicolaus.*

Bottom Row: (left to right) — *Jacob, Anna, Theresa, Mary, Henry, Michael, Jr.*

John Gehl's grandfather, Mathias Gehl, immigrates in 1845 from Keispelt, Luxembourg, and became one of the first forgers and wagonmakers.

Four of their twelve children, Nicolaus, Johann, Michael, and Henrich, become the co-founders of Gehl Brothers Manufacturing Company.

The Gehl name is spelled "Geehl" until 1892. It means "yellow" in Flemish.

John W. Gehl

Louis Lucas, a coppersmith and tin-plate worker, settles in the West Bend, Wisconsin area around 1859. He builds a foundry, working to supply the area with farm implements and machinery. In 1902, John Gehl buys the business with Henry Thoma and Peter Berres from Young and Silberzahn.

Gehl Brothers Manufacturing Company Management Team 1912

seated from left to right: John W. Gehl, Henry Gehl, Charles Silberzahn; seated in center: Mike Gehl and Nick Gehl.

1880

The Hexelbank

The first Hexelbank is produced in 1880. It is the predecessor to the modern forage harvester, cutting silage for livestock farmers from 1880 to 1906. At this time a new Hexelbank sells for about $11.50.

The 1921 Gehl sales team

The Gehl name is spread coast-to-coast by these hardworking salesmen.

The popular 1921 **Gehl silo filler**. Many companies copy the Gehl design, making flywheel-cut silo fillers the popular design of the 1920s and '30s.

1927 Manure Spreader

It is a 70-bushel model featuring knee-action and auto-type steering.